Protein Folding Kinetics

Bengt Nölting

Protein Folding Kinetics
Biophysical Methods

Second Edition
With 170 Figures, 12 in Color and 15 Tables

 Springer

Dr. Bengt Nölting
Prussian Private Institute
of Technology at Berlin
Am Schlosspark 30
D-13187 Berlin
Germany
nolting@pitb.de

Library of Congress Control Number: 2005929411

ı○○So930407

ISBN-10 3-540-27277-1 2nd Edition Springer Berlin Heidelberg New York
ISBN-13 978-3-540-27277-9 2nd Edition Springer Berlin Heidelberg New York
2nd edition 2006. Revised and extended.

ISBN 3-540-65743-6 1st Edition Springer Berlin Heidelberg New York

Springer is a part of Springer Science+Business Media
springeronline.com

© Springer-Verlag Berlin Heidelberg 1999, 2006
Printed in Germany

Typesetting: By the Author
Production: LE-TeX, Jelonek, Schmidt & Vöckler GbR, Leipzig
Coverdesign: design&production, Heidelberg

Printed on acid-free paper   2/YL – 5 4 3 2 1 0

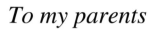

To my parents

Faust

Then shall I see, with vision clear,
How   secret   elements   cohere,
And   what   the   universe   engirds,
And give up huckstering with words.

Johann Wolfgang von Goethe

Preface

This second edition contains three new chapters covering (a) the high resolution of the folding pathways of six proteins by using the powerful method of Φ-value analysis (Chap. 11; Nölting and Andert, 2000), (b) the structural determinants of protein folding kinetics (Chap. 12; Nölting 2003; Nölting et al., 2003), and finally (c) presenting a novel method called "evolutionary computer programming" (Chap. 13; Nölting et al., 2004). The latter method involves the self-evolution of computer programs that can lead to highly advanced programs which are able to calculate protein folding and structure with unprecedented efficiency. The scope of such self-evolving computer programs is far beyond protein folding and biophysics. Section 13.3 outlines some possible future applications of self-evolving computer programs which can yield systems smarter than humans in fulfilling certain technological tasks. For further information on biophysics methods in general, the reader may refer also to the textbook "Methods in Modern Biophysics".

Mai 2005   Bengt Nölting

Preface to the first edition

The study of fast protein folding reactions has significantly advanced, following the recent development of new biophysical methods which enable not only kinetic resolution in the submillisecond time scale but also higher structural resolution. The pathways and structures of early folding events and the transition state structures of fast folding proteins can now be studied in far more detail. The validity of different models of protein folding for those events may now be elucidated and the high speed of complicated folding reactions far better understood.

This book, which is based to a high degree on several publications by the author and coworkers (e.g., Nölting, 1991, 1995, 1996, 1998a, b, 1999; Nölting et al., 1992, 1993, 1995, 1997a, b; Nölting and Sligar, 1993; Pfeil et al., 1993a, b), is particularly dedicated to students of biophysics, biochemistry, biotechnology, and medicine as a practical introduction to the modern biophysical methods of high kinetic (Chaps. 3–6, 8–10) and structural (Chaps. 2–3, 7–10) resolution of reactions that involve proteins with emphasis on protein folding reactions. Many methods are of truly interdisciplinary nature, ranging from mathematics to biophysics to molecular biology and can hardly be found in other textbooks. Since there is a rapid ongoing progress in the development and application of these methods, in particular in protein engineering, ultrafast mixing, temperature-jumping, optical triggers of folding, and Φ-value analysis, a large amount of essential information concerning the equipment and experimental details is included.

Chapter 10 reports the first high resolution of the folding pathway of a protein from microseconds to seconds (Nölting et al., 1995, 1997a, b; Nölting, 1998a). Requisite for this work was the development of a new method for the initiation and study of rapid folding which involves temperature-jumping of a set of suitably engineered mutants from the cold-unfolded to the folded state (Nölting et al., 1995, 1997a; Nölting, 1996). This new method allows fast processes that would normally be hidden in kinetic studies to be revealed.

Of course, the range of applicability of fast kinetic methods is far wider than that presented. Thus, everybody working in the fields of fast chemical reactions and physical changes, such as conformational isomerizations, enzyme kinetics and enzyme mechanisms, might see the book as a useful introduction.

The framework that is provided for the readers is the notion that the quantitation of kinetic rate constants and the visualization of protein structures along the folding pathway will lead to an understanding of function and mechanism and will aid the understanding of important biological processes and disease states through detailed mechanistic knowledge. Numerous figures provide useful information not easily found elsewhere, and the book includes copious references to original research papers, relevant reviews and monographs.

My work at Cambridge University and the Medical Research Council was supported by a European Union Human Capital and Mobility Fellowship and a Medical Research Council Fellowship. I gratefully acknowledge Prof. Dr. Alan R. Fersht for the interest in our work on fast folding reactions. NMR measurements on peptides of barstar were done by Dr. José L. Neira and Dr. Andrés S. Soler-González.

The work at the University of Illinois at Urbana-Champaign was supported by NIH grant GM31756. Prof. Dr. Steven G. Sligar is particularly acknowledged for his support of acoustic relaxation experiments and many fruitful discussions. Prof. Dr. Martin Gruebele kindly presented a LASER T-jump spectrometer with real-time fluorescence detection in the nanosecond time scale.

Dr. Robert Clegg is acknowledged for the demonstration of an ultrafast mixing device. Prof. Dr. Manfred Eigen and Dr. Dietmar Porschke kindly demonstrated a T-jump and electric field-jump apparatus. I am indebted to Dr. Min Jiang and Dr. Gisbert Berger for proof-reading the manuscript, and to Dr. Marion Hertel and Ms. Janet Sterritt-Brunner for processing the manuscript within Springer-Verlag.

Legal remarks: A number of methods mentioned in this book are covered by patents. Nothing in this publication should be construed as an authorization or implicit license to practice methods covered by any patents.

January 1999   Bengt Nölting

Contents

Symbols

Å	angström (10^{-10} m)	k_{obs}	observed rate constant, relaxation constant, decay constant (k_{obs} is denoted with "λ" in Chap. 4)
Ala	alanine		
Arg	arginine		
Asn	asparagine	L	liter
Asp	aspartic acid	Leu	leucine
bar	10^5 Pa ($= 10^5$ N m^{-2})	Lys	lysine
BdpA	B-domain of protein A from *Staphylococcus aureus*	MCD	magnetic circular dichroism
		MCT	mercury cadmium telluride
BP	base pair	Met	methionine
°C	degree Celsius (= degree Kelvin − 273.15)	MG	molten globule
		µm	micrometer (10^{-6} m)
CCD	charge-coupled device	µs	microsecond (10^{-6} s)
CD	circular dichroism	mL	milliliter
CO	carbon monoxide	mol	6.0221×10^{23}
CO	contact order	nm	nanometer (10^{-9} m)
CTP	chain topology parameter	NMR	nuclear magnetic resonance
Cys	cysteine	NOE	nuclear Overhauser effect
[D]	concentration of denaturant	ns	nanosecond (10^{-9} s)
Da	dalton, g mol^{-1}	OPO	optical parametric oscillator
ΔG_{F-U}	Gibbs free energy change upon folding	PCR	polymerase chain reaction
		Phe	phenylalanine
ΔG_{U-F}	Gibbs free energy change upon unfolding ($= -\Delta G_{F-U}$)	pI	isoelectric point
		ppm	part per million, 10^{-6}
DNA	deoxyribonucleic acid	Pro	proline
En-HD	Engrailed homeodomain	ps	picosecond (10^{-12} s)
F	folded state	PVC	polyvinyl chloride
FID	free induction decay	*R*	molar gas constant (8.3145 J mol^{-1} K^{-1})
Gln	glutamine		
Glu	glutamic acid	Ser	serine
Gly	glycine	SHG	second harmonic generation
GuHCl	guanidine hydrochloride	TFE	trifluoroethanol
h	Planck constant (6.6261×10^{-34} J s)	Thr	threonine
		T-jump	temperature-jump
His	histidine	TMS	tetramethylsilane
HPLC	high-performance liquid chromatography	Tris-HCl	tris(hydroxymethyl) aminomethane hydrochloride
		Trp	tryptophan
I	intermediate state	TY	tryptone−yeast
Ile	isoleucine	Tyr	tyrosine
IPTG	isopropyl-1-thio-β-galactoside	U	unfolded state
J	Joule (1 J = 1 Ws = 0.239 cal)	UV	ultraviolet
k_B	Boltzmann constant (1.3807×10^{-23} J K^{-1})	Val	valine
		VIS	visible
kcal	kilocalorie (= 4.18 kJ)	#	transition state
kDa	kilodalton, kg mol^{-1}		

1 Introduction

A requisite for the further understanding of the protein folding problem is the high structural and kinetic resolution of the folding pathway in the time scale from microseconds to seconds (Fersht et al., 1994; Chan, 1995; Nölting et al., 1995, 1997a, 2003, 2004; Shakhnovich et al., 1996; Wolynes et al., 1996; Eaton et al., 1996a; Dill and Chan, 1997; Nölting, 1998a, 1999, 2003; Nölting and Andert, 2000).

After decades of research, the folding mystery slowly unfolds. The amazing efficiency of the folding reaction becomes immediately obvious if one tries to imagine the huge number of conformations in the unfolded state: Estimates of the number of conformations in the maximally unfolded state, the so-called "random coil state", are around 10^{100} for a protein of 100 amino acid residues (Fig. 1.1; Finkelstein, 1997). In the folding reaction, the unique native conformation is attained on a time scale of typically seconds or even milliseconds for small proteins at room temperature, if there are no complications from slowly isomerizing amino acid residues, in particular prolines.

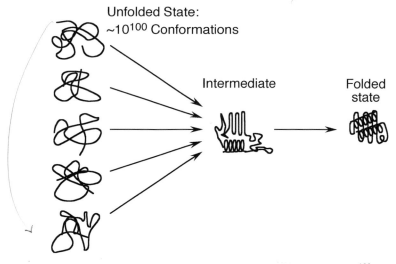

Unfolded State:
~10^{100} Conformations

Intermediate

Folded state

Fig. 1.1. Folding paradox. A protein with 100 residues may attain roughly 10^{100} conformations in the maximally unfolded state, the so-called "random coil state". Randomly sampling of all of these conformations would take many millions of years even if every sampling step would take only 1 ns. In contrast, many small proteins fold in seconds or faster. What makes the folding reaction so rapid compared to a random walk? What is the clever mechanism that has evolved? How are conformations so efficiently directed? What are the pathways of folding?

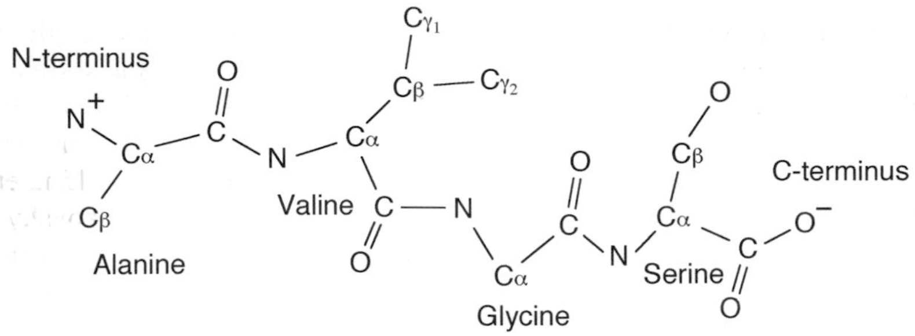

Fig. 1.2. Gigantic number of conformations of unfolded protein: In the random coil state usually several approximately independent conformations per amino acid residue are possible. The scheme of an arbitrary conformation of the polypeptide AVGS is shown. For reasons of simplicity, the hydrogen atoms are not shown.

The main reason for the gigantic number of unfolded conformations is that often the energy differences between different rotamers are small, and thus, there are comparable occupancies of many different orientations of the protein backbone and sidechains (Fig. 1.2). On average, the number of independent conformations per amino acid residue is about 10 (Finkelstein, 1997). Levinthal realized already in 1968 that folding cannot proceed via a random sampling of conformations since, even if one assumes one nanosecond per sampling step, the time of folding would be far greater than the measured time. Consequently there must be folding pathways which allow folding to proceed far more efficiently than on a random walk (Levinthal, 1968).

Several reasons have contributed to the paramount current and still increasing theoretical and experimental interest in protein folding: 1. Protein misfolding, aggregation and fibrillogenesis is connected with a number of diseases, such as prion-, Huntington's-, and Alzheimer's diseases (Bychkova and Ptitsyn, 1995; Eigen, 1996; Booth et al., 1997; Masters and Beyreuther, 1997). 2. There is a significant interest in the overexpression of recombinant proteins with the correct fold for industrial and research applications. 3. Enzymatic activity under severe conditions, such as in organic co-solvent solutions, is seen as a potentially new method for chemical synthesis (Klibanov, 1989, 1997; Griebenow and Klibanov, 1997; Kunugi et al., 1997; Wangikar et al., 1997). 4. Further, the folding problem is connected with the significant mathematical problem of finding global minima in highly complex energy-potential surfaces (Fig. 1.3) in high-dimensional spaces (Stouten et al., 1993; Luthardt and Frömmel, 1994; Cvijovic and Klinowski, 1995; Scheraga, 1996; Becker and Karplus, 1997).

Computer simulations suggest that the energy landscape along the folding pathway of proteins is often not perfectly smooth and that stable or unstable intermediates may be passed through (Itzhaki et al., 1994; Ptitsyn, 1994; Sosnick et al., 1994; Abkevich et al., 1994a; Bryngelson et al., 1995; Karplus and Sali, 1995; Onuchic et al., 1995; Baldwin, 1996; Privalov, 1996; Roder and Colon,

1997; Nath and Udgaonkar, 1997a). Especially, proteins in which a single, very deep global energy minimum is absent may display poor foldability and complicated pathways with a number of early intermediates (Fersht, 1995c; Abkevich et al., 1996; Shakhnovich, 1997). In particular, so-called molten globule intermediates have found significant attention (Dolgikh et al., 1981; Nölting et al., 1993; Ptitsyn, 1994, 1995; Chalikian et al., 1995; Fink, 1995; Gussakovsky and Haas, 1995; Kuwajima, 1996; Fink et al., 1998). Another source of the occurrence of intermediates is the existence of co-factors which often have dramatic contributions to protein stability (Pfeil, 1981, 1993; Pfeil et al., 1991, 1993a, b; Elöve et al., 1994; Burova et al., 1995).

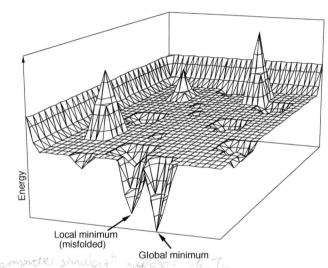

Local minimum
(misfolded)

Global minimum

Fig. 1.3. Energy landscape of a protein. Only two reaction coordinates can be drawn. In reality the energy landscape represents an n-dimensional hyper-surface in the $(n+1)$-dimensional space, where n is the degree of freedom of conformational movement of the molecule.

On the other hand and surprisingly, small proteins have been discovered which may complete the whole folding reaction in the submillisecond time scale (Khorasanizadeh et al., 1993; Schindler et al., 1995; Robinson and Sauer, 1996; Sosnick et al., 1996; Chan et al., 1997; Takahashi et al., 1997). Fast folding sequences are found far easier if the structure of the protein is symmetric (Wolynes et al., 1995; Wolynes, 1996). The maximum rate for protein folding is estimated to be of the order of 1 μs^{-1} (Hagen et al., 1996, 1997) !

The occurrence of rapid events in the submillisecond time scale has been detected indirectly with slow methods by the observation of burst-phases, i.e., changes of the signal within the dead time of the method (Fig. 1.4). However, the precise and comprehensive analysis of early events requires a direct kinetic resolution. In the past years we have seen a remarkable progress in the develop-

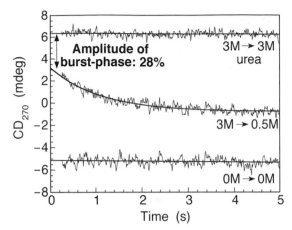

Fig. 1.4. Burst-phase observed when refolding the 10 kDa protein C40A/C82A/P27A barstar at 5°C. Jumps are with different concentrations of urea as indicated. In 3 M urea the protein is more than 80% unfolded, and it is more than 95% folded in 0 M urea. The circular dichroism (CD) at 270 nm mainly reflects the structure consolidation in the vicinity of the 8 aromatic amino acid residues.

ment of new methods which enable us to access the submillisecond, microsecond, and even nanosecond time scale of protein folding (Fig. 1.5). The high kinetic and structural resolution has profoundly altered the picture of folding reactions and enhanced the understanding of the tremendous speed and efficiency of protein folding (Nölting et al., 1995, 1997, 2003; Plaxco and Dobson, 1996; Wolynes et al., 1996; Eaton et al., 1996a, 1997; Nölting and Andert, 2000). This book focuses on the biophysical principles of the kinetic methods and the high structural resolution of folding.

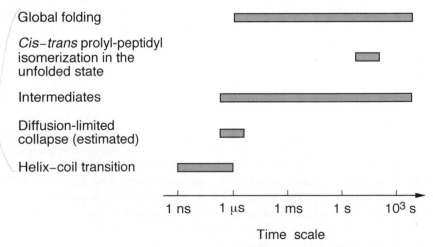

Fig. 1.5. Typical time scale of folding events under standard conditions, 25°C, pH 7.

2 Structures of proteins

2.1
Primary structure

The main building blocks of proteins are residues of 20 natural α-amino acids. With the exception of proline, their structure is:

$$
\begin{array}{c}
R \\
| \\
H_2N \longrightarrow CH \longrightarrow CO_2H
\end{array}
\tag{2.1}
$$

In proline the sidechain, R, is bridged to the nitrogen atom of the amino group. The structures of the sidechains and the properties of the amino acids are given in Tables 2.1 and 2.2, respectively. The amino acid sidechains can be grouped in the categories given in Table 2.2 or, alternatively, in the categories non-polar (glycine, alanine, valine, leucine, isoleucine, proline, phenylalanine), polar (serine, threonine, cysteine, methionine, asparagine, glutamine, tyrosine, tryptophan), and charged (aspartic acid, glutamic acid, lysine, arginine, histidine).

With the exception of glycine, which is not chiral, all acids of natural occurring proteins are L-isomers and are optically active. Tryptophan, tyrosine and phenylalanine absorb light at wavelengths below 310 nm, 300 nm and 270 nm, respectively. The first absorption maximum is around 280 nm for tryptophan and tyrosine and around 260 nm for phenylalanine. At 280 nm, the absorption of tyrosine is 4 times lower than that of tryptophan at pH 6. Phenylalanine absorption at 260 nm is 6 times lower than tyrosine absorption at 280 nm (Wetlaufer, 1962).

In proteins, the amino acids are linked together by the peptide bond (Eq. 2.2), which is formed upon condensation of two amino acids (Creighton, 1993).

$$
\begin{array}{c}
R_1 \qquad\qquad\qquad\qquad R_2 \\
| \qquad\qquad\qquad\qquad | \\
H_2N \longrightarrow CH \longrightarrow CO_2H \;+\; H_2N \longrightarrow CH \longrightarrow CO_2H \\
\downarrow \\
R_1 \quad O \qquad\qquad R_2 \\
| \quad || \qquad\qquad | \\
H_2N \longrightarrow CH \longrightarrow C \longrightarrow NH \longrightarrow CH \longrightarrow CO_2H \;+\; H_2O \\
\uparrow \\
\text{Peptide bond}
\end{array}
\tag{2.2}
$$

Table 2.1. Structure of the sidechains, R, of natural amino acids. For proline, the backbone nitrogen and C_αH-group are included.

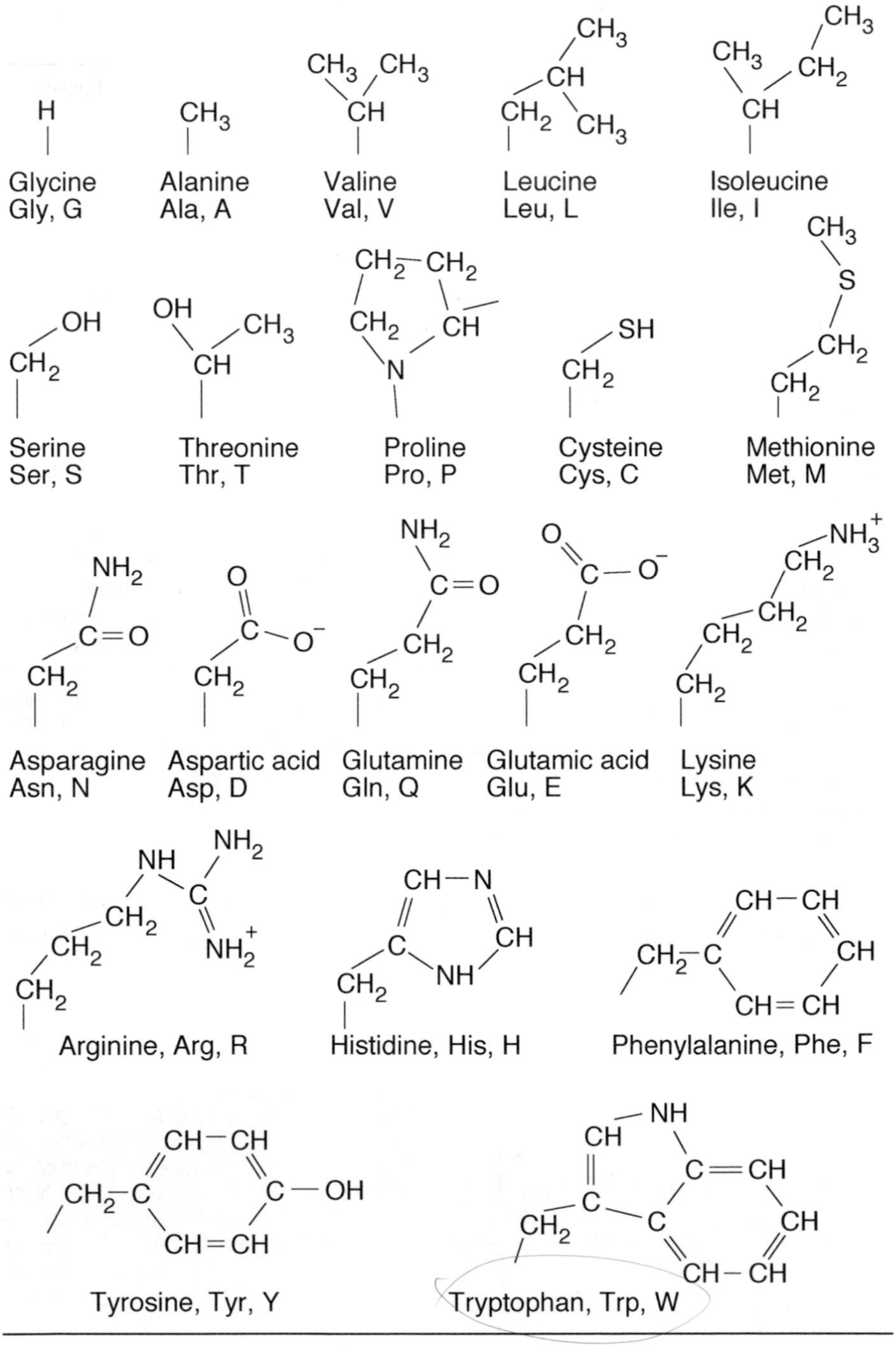

H	CH_3	CH_3 CH_3 / CH	CH_3 CH / CH_2 CH_3	CH_3 CH_2 / CH_3 CH
Glycine Gly, G	Alanine Ala, A	Valine Val, V	Leucine Leu, L	Isoleucine Ile, I

Serine, Ser, S — OH, CH_2

Threonine, Thr, T — OH, CH, CH_3

Proline, Pro, P — CH_2–CH_2, CH_2, CH, N

Cysteine, Cys, C — SH, CH_2

Methionine, Met, M — CH_3, S, CH_2, CH_2

Asparagine Asn, N — NH_2, C=O, CH_2

Aspartic acid Asp, D — O, C–O^-, CH_2

Glutamine Gln, Q — NH_2, C=O, CH_2, CH_2

Glutamic acid Glu, E — O, C–O^-, CH_2, CH_2

Lysine Lys, K — NH_3^+, CH_2, CH_2, CH_2, CH_2

Arginine, Arg, R — NH, NH_2, C, NH_2^+, CH_2, CH_2, CH_2

Histidine, His, H — CH–N, C, CH, CH_2, NH

Phenylalanine, Phe, F — CH–CH, CH_2–C, CH, CH=CH

Tyrosine, Tyr, Y — CH–CH, CH_2–C, C–OH, CH=CH

Tryptophan, Trp, W — NH, CH, C=CH, CH_2–C, C, CH, CH–CH

Table 2.2. Properties of amino acid residues commonly found in proteins.

Amino acid	Residue mass[a] (daltons)	Van der Waals volume[b] ($Å^3$)	Frequency in proteins[c] (%)	Chemical type of sidechain
Alanine	71.08	67	8	inactive, aliphatic
Arginine	156.19	148	6	basic
Asparagine	114.10	96	4	amide
Aspartic acid	115.09	91	5	acidic
Cysteine	103.14	86	2	sulfur-containing
Glutamic acid	129.12	109	6	acidic
Glutamine	128.13	114	4	amide
Glycine	57.05	48	7	inactive
Histidine	137.14	118	2	basic
Isoleucine	113.16	124	5	inactive, aliphatic
Leucine	113.16	124	9	inactive, aliphatic
Lysine	128.17	135	6	basic
Methionine	131.20	124	2	sulfur-containing
Phenylalanine	147.18	135	4	inactive, aromatic
Proline	97.12	90	5	inactive
Serine	87.08	73	7	hydroxyl
Threonine	101.11	93	6	hydroxyl
Tryptophan	186.21	163	1	indole, aromatic
Tyrosine	163.18	141	3	hydroxyl, aromatic
Valine	99.13	105	7	inactive, aliphatic

For the calculation of the molecular weight of proteins or peptides, 18 daltons have to be added for the −H and −OH at N- and C-termini, respectively. 2 daltons are to be subtracted per disulfide bridge. The sidechains labeled with "inactive" react only under extreme conditions.
[a] For neutralized sidechains (Lide, 1993; Creighton, 1993; Coligan et al., 1996).
[b] (Richards, 1974; Creighton, 1993).
[c] (McCaldon and Argos, 1988; Creighton, 1993; Coligan et al., 1996).

The short distance of the peptide bond, C–N (Eq. 2.2; Fig. 2.1), of only about 1.31–1.34 Å, compared with about 1.44–1.48 Å for the nonpeptide C–N bonds, reflects its partial double-bond character.

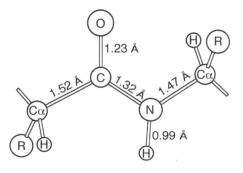

Fig. 2.1. The peptide bond. Distances are averages observed in several protein NMR and crystal structures (Brookhaven National Laboratory Protein Data Bank; Abola et al., 1987, 1997).

Due to the partial double-bond character of the peptide bond (Eq. 2.3), the adjacent atom groups have a strong tendency to be coplanar.

$$
\begin{array}{c}
\underset{\text{C}\alpha}{\overset{\text{O}}{\underset{\diagup}{\overset{\parallel}{\text{C}}}}}\!\!-\!\!\underset{\text{H}}{\overset{\text{C}\alpha}{\underset{\diagup}{\text{N}}}}
\quad\rightleftharpoons\quad
\underset{\text{C}\alpha}{\overset{\text{O}^{-}}{\underset{\diagup}{\text{C}}}}\!\!=\!\!\underset{\text{H}}{\overset{\text{C}\alpha}{\overset{+}{\text{N}}}}
\end{array}
\qquad (2.3)
$$

$$
\begin{array}{c}
\underset{\text{C}\alpha}{\overset{\text{O}}{\underset{\diagup}{\overset{\parallel}{\text{C}}}}}\!\!-\!\!\underset{\text{C}\alpha}{\overset{\text{H}}{\underset{\diagup}{\text{N}}}}
\quad\rightleftharpoons\quad
\underset{\text{C}\alpha}{\overset{\text{O}}{\underset{\diagup}{\overset{\parallel}{\text{C}}}}}\!\!-\!\!\underset{\text{H}}{\overset{\text{C}\alpha}{\underset{\diagup}{\text{N}}}}
\end{array}
\qquad (2.4)
$$

$\quad\quad\quad\quad\quad$ *cis* $\quad\quad\quad\quad\quad\quad\quad\quad$ *trans*

Two configurations of the peptide bond are possible, the *trans*- and the *cis*-form (Eq. 2.4). Because of steric hindrance between adjacent sidechains in the *cis*-form, the *trans*-form is usually energetically favored by more than 17 kJ mol^{-1} (4 kcal mol^{-1}), corresponding to less than 0.1% occupancy of the *cis*-isomer. For the prolyl-peptidyl bond, the energy difference is significantly reduced: in small peptides it is typically only about 2–3 kJ mol^{-1} (0.5–0.7 kcal mol^{-1}), corresponding to about 70–80% population of the *trans*-configuration (Fersht, 1985; Schreiber, 1993b; Creighton, 1993).

The pK$_a$'s found in single amino acids change upon incorporation into the protein due to the change of environment (Table 2.3). The acidic residues of aspartic acid and glutamic acid are negatively charged and the basic residues of lysine and arginine have a positive charge at pH 7. Histidine, which has a pK$_a$ 6–7, is a strong base at neutral pH and is involved in many enzymatic reactions that involve a proton transfer.

Table 2.3. Observed pK$_a$'s of ionizable groups, found for single amino acids and for amino acid residues in proteins.

Ionizable group	pK$_a$ of amino acids[a]	pK$_a$ of amino acid residues in proteins[b]
α-Carboxyl	1.8 – 2.4	3.5 – 4.3
β-Carboxyl (aspartic acid)	3.8	3.9 – 4.0
γ-Carboxyl (glutamic acid)	4.1	4.3 – 4.5
Imidazole (histidine)	6.0	6.0 – 7.0
α-Amino	8.8 – 10.6	6.8 – 8.0
Thiol (cysteine)	8.3	9.0 – 9.5
Phenolic hydroxyl (tyrosine)	10.1	10.0 – 10.3
ε-Amino (lysine)	10.7	10.4 – 11.1
δ-Guanido (arginine)	12.5	12.0

[a] (Dawson et al., 1969; Fersht, 1985; Zubay, 1993; Lide, 1993).
[b] (Bundi and Wüthrich, 1979; Matthew, 1985; Creighton, 1993; Coligan et al., 1996).

2.2
Secondary structure

Three main elements of well-defined secondary structure may be distinguished: the α-helix, the β-sheet, and turns. These structural elements may be connected with each other by loops. Helices are the most abundant form of secondary structure in globular proteins, followed by sheets, and in the third place turns. Secondary structure formation provides an efficient mechanism of pairing polar groups of the polypeptide backbone by hydrogen bonds. Uncoupling of only a single pair of polar groups in the protein interior may cause an energy cost of 40 kJ mol^{-1} (9.6 kcal mol^{-1}) (Privalov and Makhatadze, 1993). Protein secondary structure can directly be observed by atomic force microscopy (McMaster et al., 1996). Secondary structure elements may associate through sidechain interactions to form super-secondary structure, so-called motifs.

Different probabilities are observed for the incorporation of amino acid residues into different types of secondary structure (O'Neil and DeGrado, 1990; Creighton, 1993; Coligan et al., 1996; Hubbard et al., 1996). Using X-ray crystallographic data of a large set of proteins, Chou and Fasman (Chou and Fasman, 1977, 1978a, 1978b; Chou, 1989) calculated statistical conformational preference parameters which were based on the occurrence of a specific amino acid type in a specific type of secondary structure, on the relative frequency of that amino acid type in the databases, and on the relative number of amino acid residues occurring in each type of secondary structure (Table 2.4). For example, prolines and glycines are considered as helix-breakers since their preference parameters for helices are less than half of that of alanine, a so-called helix-former. Further progress has been made with the recognition that the conformational preference depends on the relative position in the secondary structure element (Presta and Rose, 1988; Richardson and Richardson, 1988; Harper and Rose, 1993). For example, glycine, serine, and threonine often constitute the amino-terminal residues (N-cap) in α-helices. Glycine and asparagine are frequently found at the carboxyl-terminal position (C-cap) of α-helices. They are referred to as being N-cap and C-cap stabilizers, respectively.

The α-helix (Fig. 2.2) is stabilized by hydrogen bonds between the carbonyl oxygen of the amino acid residue at the position n in the polypeptide chain with the amide group, NH, of the residue $n + 4$.

Hydrogen bonds between carbonyl oxygens and amide groups of adjacent strands stabilize β-sheets (Fig. 2.3). The occurrence of β-sheets is often correlated with high hydrophobicities (see Sect. 3.4) of the involved amino acid residues: Isoleucine, valine, tyrosine, and phenylalanine prefer β-sheet structure, but aspartic acid and glutamic acid have an aversion to incorporation into β-sheets (Table 2.4).

Turns (Fig. 2.4) involve a 180° change in direction of the polypeptide chain and are stabilized by a hydrogen bond between the carbonyl oxygen of the residue at the position n with the amide group, NH, of the residue $n + 3$ (Fersht, 1985).

Less common elements of secondary structure are also 3_{10}-helices and hairpins.

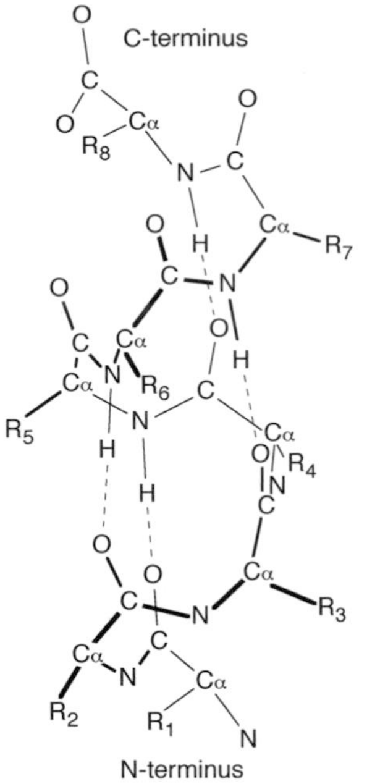

O

C-terminus

Fig. 2.2. Right-handed α-helix consisting of 8 amino acid residues. α-Helices are stabilized by hydrogen bonds between the carbonyl oxygen atom of amino acid residue number n and the amide group, NH, of residue number $n + 4$ in the polypeptide chain, as indicated by dashed lines (for simplicity, only 4 of the hydrogen atoms are shown).

N-terminus

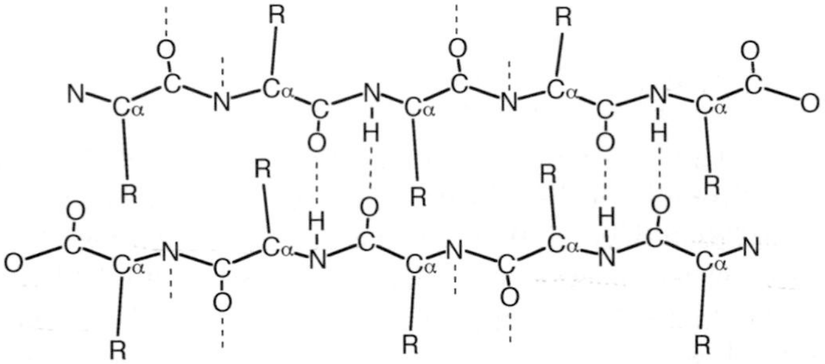

Fig. 2.3. Antiparallel β-sheet consisting of 10 amino acid residues. The β-sheet is stabilized by hydrogen bonds between carbonyl oxygen atoms and amide groups, NH, of adjacent strands, as indicated by dashed lines (for simplicity, only 4 of the hydrogen atoms are shown).

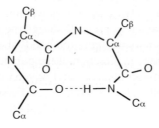

Fig. 2.4. Type I turn. Turns are stabilized by a hydrogen bond between amino acid residues n and $n + 3$, as indicated by a dashed line (for simplicity, the other hydrogen atoms are not displayed).

Table 2.4. Preferences of amino acids for different types of secondary structure.

Amino acid	Conformational preference parameter[a]		
	α-Helix	β-Strand	Turn
Alanine	1.3 – 1.5	0.8 – 0.9	0.7
Arginine	0.9 – 1.4	0.7 – 1.0	0.9 – 1.0
Asparagine	0.8 – 0.9	0.6 – 0.7	1.3 – 1.6
Aspartic acid	0.9 – 1.1	0.5 – 0.7	1.4 – 1.5
Cysteine	0.9 – 1.0	0.8 – 1.2	0.9 – 1.2
Glutamic acid	1.4	0.5 – 0.8	0.7 – 1.0
Glutamine	1.1 – 1.4	0.8 – 1.0	1.0
Glycine	0.4 – 0.6	0.6 – 0.9	1.6
Histidine	1.0 – 1.2	0.8 – 1.1	0.7 – 1.0
Isoleucine	1.0 – 1.1	1.5 – 1.8	0.5
Leucine	1.3	1.0 – 1.2	0.6
Lysine	1.1 – 1.2	0.7 – 0.9	1.0
Methionine	1.3 – 1.4	1.0 – 1.3	0.4 – 0.6
Phenylalanine	1.0 – 1.1	1.2 – 1.4	0.6
Proline	0.5 – 0.6	0.4 – 0.6	1.5 – 1.9
Serine	0.7 – 0.8	0.9 – 1.0	1.3 – 1.4
Threonine	0.7 – 0.8	1.2 – 1.3	1.0
Tryptophan	1.0	1.2	0.8 – 1.0
Tyrosine	0.7 – 0.9	1.2 – 1.5	1.1
Valine	0.9 – 1.0	1.5 – 1.7	0.5

[a] (Chou and Fasman, 1977, 1978a, b; Chou, 1989; Creighton, 1993; Thornton et al., 1995).

2.3
Tertiary structure

The polypeptide chain of natural proteins is unbranched but may contain disulfide bridges. Some proteins can incorporate cofactors, e.g., heme or chlorophyll. Protein cores are packed as tightly as organic solids, or slightly tighter. Also the compressibility of the interior of globular proteins, $(14 \pm 3) \times 10^{-11}$ Pa^{-1}, is comparable with that of soft organic solids, and is 3 times less than the compressibility of water. Partial volumes of globular proteins in aqueous solution typically are $\approx 0.70-0.75$ L kg^{-1} (Sarvazyan, 1991; Kharakoz and Sarvazyan, 1993; Nölting and Sligar, 1993).

For most proteins the tertiary structure is very well defined, and many of the sidechain rotation-, segmental flexibility-, and molecular breathing motions are on a scale of less than 2 Å. Sidechains located in the interior of the protein molecule usually rotate or perform 180° flips with frequencies of 10^2-10^7 Hz. Rotation of buried tryptophan sidechains is usually so infrequent that they are considered as almost immobile. However, structural transitions which involve large conformational changes can play a crucial role in enzymatic and binding reactions. In the crystal structures of some proteins whole domains are statically disordered, i.e., different conformations are occupied.

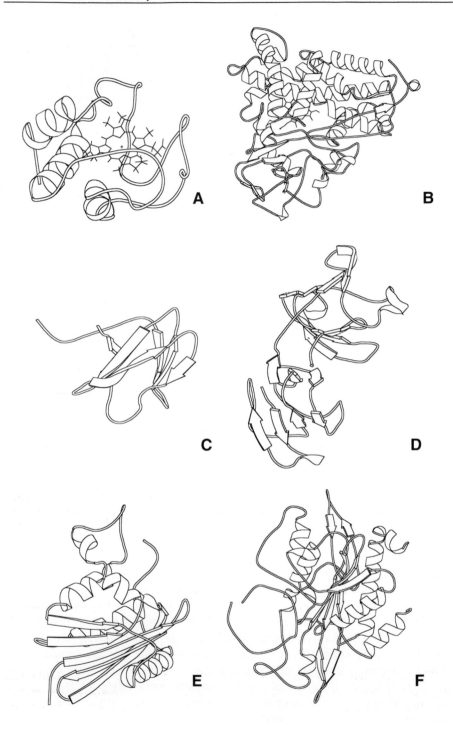

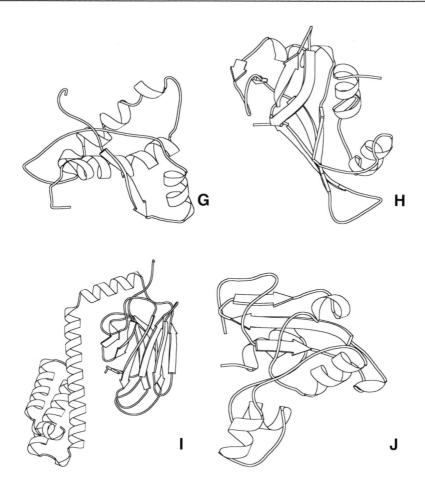

Fig. 2.5. Examples for classes of folds of soluble proteins (Poulos et al., 1986; Abola et al., 1987, 1997; Billeter et al., 1990; Kim et al., 1990; Katayanagi et al., 1992; Tilton et al., 1992; Korolev et al., 1995; Murzin et al., 1995; Kumaraswamy et al., 1996; Qi et al., 1996; Riek et al., 1996; Zhu et al., 1996; Chothia et al., 1997). The protein backbones are symbolized by ribbons. The heme cofactors of cytochrome c and cytochrome P-450cam are shown as wireframes. All alpha (or almost all alpha): A (cytochrome c), B (cytochrome P-450cam). All beta: C (α-amylase inhibitor), D (γ-crystallin). Alpha and beta (parallel or antiparallel β-sheets; non-segregated or segregated α- and β-regions): E (ribonuclease H), F (restriction endonuclease EcoRI bound to DNA), G (prion protein domain), H (ribonuclease A). Multi domain: I (substrate-binding domain of the chaperone DnaK), J (fragment of *Thermus aquaticus* DNA polymerase). The figure was drawn using the program MOLSCRIPT (Kraulis, 1991).

The unique tertiary structure of each protein is determined by its amino acid sequence. Usually, the native structure of protein is at the minimum free energy (see Fig. 1.3). Exceptions to this basic tenet of protein folding are very rare (Sohl et al., 1998).

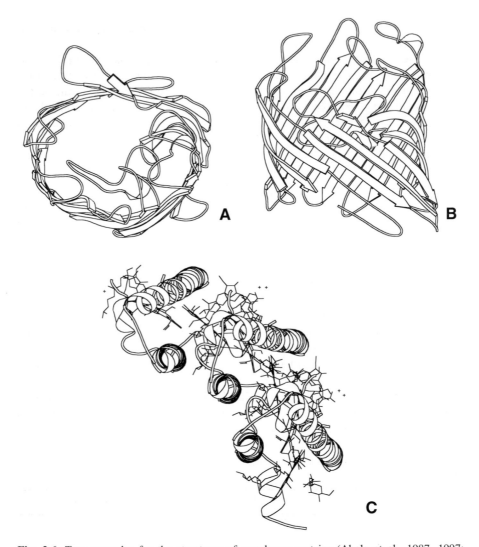

Fig. 2.6. Two examples for the structures of membrane proteins (Abola et al., 1987, 1997; Cowan et al., 1992; Murzin et al., 1995; Chothia et al., 1997; Prince et al., 1997). A, B: phosphoporin, C: fragment of the purple bacteria light-harvesting complex LH2. The backbones are symbolized by ribbons and the cofactors chlorophyll and carotenoid of the light-harvesting protein are shown as wireframes. The figure was drawn using the program MOLSCRIPT (Kraulis, 1991).

Nature has evolved a gigantic variety of protein three-dimensional structures, so-called protein folds. For the thousands of coordinate files deposited in protein databases, most notably the Brookhaven National Laboratory Protein Data Bank (Abola et al., 1987, 1997) which may be accessed via the World Wide Web with an entry point http://www.rcsb.org/pdb/, 7 classes of folds with more than 270 folds may be distinguished (Table 2.5; Murzin et al., 1995). Examples for the huge variety of protein folds are presented in Figs. 2.5 and 2.6.

Together with the chemical diversity of the sidechains, this structural variety contributes to an astonishing functional diversity of proteins, ranging from catalysis or inhibition of various chemical or physical reactions, to the transport of other molecules, electrons, protons, or excitons, to the stabilization of the architecture within cells. The examples in Fig. 2.5 are soluble proteins. Many important proteins incorporate cofactors that are covalently (e.g., heme in cytochrome c; Fig. 2.5) or non-covalently bound (e.g., heme in cytochrome P-450$_{cam}$; Fig. 2.5). Membrane proteins are exemplified by phosphoporin and a fragment of the purple bacteria light-harvesting complex LH2 (Fig. 2.6). The protein matrix of the LH2 fragment embeds 9 chlorophyll- and at least 3 carotenoid molecules. Complicated structures like this have evolved over many millions of years and contribute to an amazing efficiency of light-harvesting complexes of bacteria and higher plants, that cannot be reproduced *in vitro* when using a simple solution of chlorophyll molecules. In some higher plants, up to 98% of the absorbed photons are transmitted to the reaction centers via an exciton–exciton transfer mechanism.

The visualization of protein structures in the folded state and along the folding pathway, and the quantitation of kinetic rate constants is seen to be of paramount importance for an understanding of protein function and mechanism, and will aid the understanding of important biological processes and disease states through detailed mechanistic knowledge. The next chapters are devoted to the mathematical, biophysical, chemical, and molecular biological methods of high kinetic and structural resolution of chemical and biophysical reactions of proteins with emphasis on folding reactions.

Table 2.5. Classes of folds found in the protein databases (Murzin et al., 1995; Chothia et al., 1997).

Class of protein fold	Relative abundance
All alpha	20 – 30%
All beta	10 – 20%
Alpha and beta with mainly parallel β-sheets (α/β)	15 – 25%
Alpha and beta with mainly antiparallel β-sheets with segregated α- and β-regions (α+β)	20 – 30%
Multi domain (alpha and beta)	< 10%
Membrane and cell surface proteins	< 10%
Small proteins (dominated by cofactors or disulfide bridges)	5 – 15%

3 Physical interactions that determine the properties of proteins

This chapter gives an introduction into the physical forces that determine, together with covalent interactions, the conformations along the folding pathway, including the folded and unfolded structures. These forces also dominate the non-covalent mutual interactions between (a) two protein molecules, (b) proteins and other macromolecules, and (c) proteins and solvent. Further information may be found in Cantor and Schimmel, 1980; Fersht, 1985; Creighton, 1993; Makhatadze and Privalov, 1993; Privalov and Makhatadze, 1993.

Electrostatic interactions of point charges (Sect. 3.1.1) crucially affect most long-range interactions of proteins with proteins and other charged macromolecules. Van der Waals interactions (Sects. 3.1.2 and 3.2) are considered the main contributors to the stabilization of globular proteins, followed by hydrogen bonds (Sect. 3.3), and in the third place hydrophobic interactions (Sect. 3.4) of non-polar residues (Privalov and Makhatadze, 1993). In order to produce a stable folded protein conformation, these contributions have to overcompensate the destabilizing contributions from the hydration of polar residues (see Sect. 3.4) and the gain in configurational entropy upon unfolding. The magnitudes of stabilizing and destabilizing contributions to the overall protein stability typically are several 1000 kJ mol^{-1}. A delicate balance between stabilizing and destabilizing contributions causes a stability of most globular proteins in water in the range of only 10–70 kJ mol^{-1} (Privalov, 1979; Privalov and Makhatadze, 1993).

3.1
Electrostatic interactions

3.1.1
Point charges

Coulomb's law provides the force, F, between two charged species, Z_1 and Z_2

$$F = \frac{1}{4\pi\varepsilon_o\varepsilon_r} \frac{Z_1 Z_2}{d_{1,2}^2} \,, \tag{3.1}$$

where $d_{1,2}$ is the distance between Z_1 and Z_2, $\varepsilon_o = 8.854\times10^{-12}$ C V^{-1} m^{-1} is the permittivity of vacuum, and ε_r is the relative permittivity. The change of energy, ΔE, as function of the distance separation is obtained by integration:

$$\Delta E = \int_{d_1}^{d_2} F dd_{1,2} = \frac{Z_1 Z_2}{4\pi\varepsilon_o\varepsilon_r}(\frac{1}{d_1} - \frac{1}{d_2}) \,. \tag{3.2}$$

ε_r reflects the polarizability of the medium between the charges. It largely differs between different solvents (Table 3.1). For example, the energy necessary to separate positive and negative elementary charges, $e = 1.602 \times 10^{-19}$ C, from a distance of 10 Å to infinity in vacuum is 139 kJ mol^{-1} (33.3 kcal mol^{-1}). In water the energy is decreased by a factor of roughly $\varepsilon_r = 78$.

Solvents that correspond chemically to the interior of proteins have a relative permittivity, ε_r, which is roughly one order of magnitude lower than that of water. Thus, Coulomb interactions of charges in the interior of proteins are typically one order of magnitude stronger than at the surface of proteins in aqueous solution. For example, surface charge mutations often change the protein stability by less than 4 kJ mol^{-1} (1 kcal mol^{-1}), while changes of more than 4 kJ mol^{-1} are not unusual for buried charge mutations.

Point charges have a wide range of interaction. In the folding reaction, Coulomb interactions can effectively steer one structural element towards another distant structural element. Coulomb interactions can also steer one protein molecule towards another. For example, the positively charged active site of the ribonuclease barnase steers the negatively charged inhibitor barstar into the optimal position for binding (Schreiber et al., 1994). Strong electrostatic protein–protein interactions can result in a very strong association (Schreiber et al., 1994).

Protein–protein complexes that are stabilized mainly by electrostatic interactions, rapidly become weakened with increasing salt concentration because protein charges become neutralized by counter ions. Proteins with large net charges may often be stabilized by salts that suppress the intramolecular charge repulsion.

Table 3.1. Properties of solvents.

Solvent	Molecular formula	Relative permittivity, ε_r[a]	Hydrophilicity[b] (kJ g^{-1})
Water	H_2O	78.4	
Methanol	CH_4O	32.7	−0.67
Ethanol	C_2H_6O	24.3	−0.46
2-Propanol	C_3H_8O	18.3	−0.33
2-Propanone	C_3H_6O	20.7	−0.28
2-Butanone	C_4H_8O	18.5	−0.21
2-Pentanone	$C_5H_{10}O$	15.5	−0.17
2-Hexanone	$C_6H_{12}O$	14.6	−0.14
Phenol	C_6H_6O	9.8	−0.29
Benzene	C_6H_6	2.3	−0.05
Hexane	C_6H_{14}	1.9	0.12

[a] (Lide, 1993).
[b] Gibbs free energies of transfer form the gaseous phase into water (Cabani et al., 1981; Privalov and Makhatadze, 1993); 1 kJ g^{-1} = 0.24 kcal g^{-1}.

3.1.2
Point charge–dipole and dipole–dipole interactions

The energy of interaction of a point charge with an induced dipole (for example, of interaction of a polarizable molecule with an ion) falls off as d^{-4}, where d is the distance of separation between charge and dipole (Fersht, 1985).

The energies of interaction between (a) randomly oriented permanent dipoles, (b) a permanent dipole and a dipole induced by it, and (c) mutually induced dipoles fall off approximately as d^{-5} to d^{-6} (Fersht, 1985; Creighton, 1993). These types of interactions are the main origin of the attractive component of the "van der Waals forces" (see Figs. 3.1–3.3 in Sect. 3.2). Type (c) occurs between all atoms and is also known as the "dispersion forces" or "London forces".

3.2
Quantum-mechanical short-range repulsion

The repulsive component of the van der Waals interaction (Figs. 3.1–3.2) falls off approximately as d^{-12} to e^{-d}, where d is the distance of separation. Its main origin is the quantum-mechanical Pauli exclusion principle. Note that historically only the attractive forces (a) and (c) in Sect. 3.1.2 were called "van der Waals forces".

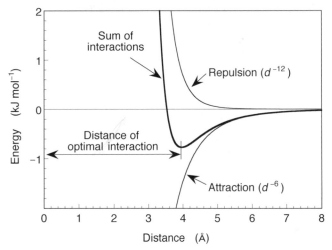

Fig. 3.1. Van der Waals potential as function of the distance separation for the interaction of two carbon atoms with $C_6 = 6\times10^3$ Å^6 kJ mol^{-1}, $C_{12} = 1.1\times10^7$ Å^{12} kJ mol^{-1} (Warshel and Levitt, 1976; Creighton, 1993). The van der Waals potential contains an attractive component that mainly originates from mutually induced dipole–dipole interactions and falls off with the sixth power of distance separation, and a repulsive component that mainly originates from the Pauli-exclusion and falls off with the twelfth power of the distance separation. Van der Waals interactions have a short range of only a few Å (1 Å = 10^{-10} m). The energies of van der Waals interactions of the atoms commonly found in proteins are small and of the order of only 0.1–2 kJ mol^{-1}, compared with the energy of 10–60 kJ mol^{-1} per hydrogen bond, and the energy of up to several 10 kJ mol^{-1} per buried salt-bridge (1 kJ mol^{-1} = 0.24 kcal mol^{-1}).

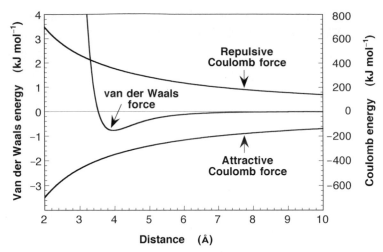

Fig. 3.2. Van der Waals potential as function of distance separation for the interaction of two carbon atoms with $C_6 = 6 \times 10^3$ Å^6 kJ mol^{-1}, and $C_{12} = 1.1 \times 10^7$ Å^{12} kJ mol^{-1}, compared with the Coulomb interaction in vacuum of two elementary charges, $e = 1.602 \times 10^{-19}$ As (1 kJ mol^{-1} = 0.24 kcal mol^{-1}). The force is repulsive for the same sign of the charges, otherwise it is attractive. Compared with Coulomb forces of point changes, van der Waals interactions are intrinsically weak and have a short range of interaction. However, cooperation of a large number of van der Waals interactions can produce a stable conformation (Creighton, 1993).

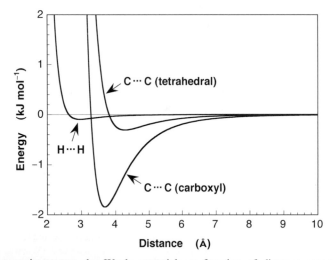

Fig. 3.3. Approximate van der Waals potentials as function of distance separation for the interaction of two hydrogen atoms, two tetrahedral carbon atoms, and two carboxyl carbon atoms, respectively, calculated using data from (Warshel and Levitt, 1976; Fersht, 1985). The van der Waals potentials of hydrogen, carbon, nitrogen, oxygen, and sulfur atoms display a shallow attractive energy minimum at distances of about 2.6–4.4 Å, corresponding to radii of 1.3–2.2 Å, and a strong repulsion at shorter distances. Van der Waals interactions of hydrogen atoms are intrinsically weaker than those of carbon atoms. Usually, carboxyl carbon atoms have a stronger interaction and a shorter van der Waals radius than tetrahedral carbon atoms.

The different components of the van der Waals interaction are often approximated by the Lennard–Jones 6,12 potential:

$$E = \frac{C_{12}}{d^{12}} - \frac{C_6}{d^6} \tag{3.3}$$

$$C_{12} = C_{12,i} \times C_{12,j} \qquad C_6 = C_{6,i} \times C_{6,j} \,,$$

where $C_{6,i}$, $C_{12,i}$, $C_{6,j}$, $C_{12,j}$ are parameters for the atoms i, and j, respectively, which usually are derived for the potential between two atoms of the same type.

Even though the van der Waals interactions are weak, in proteins they accumulate to a significant amount. The strength of the individual interaction depends on the types of interacting atoms, and varies with the chemical environment of the atoms involved. For example, for carboxyl carbon atoms the interactions are usually stronger, and van der Waals distances are usually shorter than for tetrahedral carbon atoms (Fig. 3.3).

3.3
Hydrogen bonding

A hydrogen bond contains both positive (H-donor) and negative (H-acceptor) partial charges. It represents a combination of covalent and electrostatic interactions, but the main component is the electrostatic attraction between hydrogen donor and acceptor. The magnitude of reduction of the van der Waals distance is indicative of the strength of the hydrogen bond (Table 3.2). The Gibbs free energy contributions per hydrogen bond in the interior of proteins are estimated to be $10–60$ kJ mol^{-1} ($2–14$ kcal mol^{-1}) (Hagler et al., 1979; Dauber and Hagler, 1980; Privalov and Makhatadze, 1993). Consider, for example, the hydroxyl–carbonyl bond which is one of the strongest hydrogen bonds in proteins:

$$-O-H^{\delta+} \cdots\, ^{\delta^-}O=C< \tag{3.4}$$

The electronegativity of the hydroxyl oxygen atom causes a positive partial charge of the hydroxyl hydrogen atom, the H-donor. Similarly, the carbonyl oxygen atom has a negative partial charge which attracts the hydroxyl hydrogen atom.

Table 3.2. Properties of hydroxyl–hydroxyl and amide–carbonyl hydrogen bonds found in proteins.

Type of hydrogen bond	Molecular formula	Typical H$\cdots$O distance (Å)	Typical reduction of van der Waals distance
hydroxyl – hydroxyl	$-OH\cdots OH-$	$1.9 – 2.3$	$20 – 25\%$
amide – carbonyl	$>NH\cdots O=C<$	$1.8 – 2.2$	$20 – 30\%$

Hydrogen bonding of proteins in aqueous solution is profoundly altered by addition of co-solvents. Hydrophobic co-solvents, for example, phenol and benzene, may form significantly fewer hydrogen bonds with polar and charged groups at the surface of proteins than water, and can destabilize most native proteins (see Sect. 3.4). Trifluoroethanol (TFE) stabilizes helices by strengthening their hydrogen bonds but destabilizes most native proteins by weakening the hydrophobic interaction in the core of the protein (Luo and Baldwin, 1998).

3.4
Hydrophobic interaction

Table 3.3. Properties of amino acids.

Amino acid	Codes		Accessible surface area of residue[a] (Å^2)	Relative hydrophilicity of amino acid residues[b] $\Delta G_{\text{cyclohexane}\to\text{water}}$ (kJ g^{-1})
Alanine	Ala	A	115	0.05
Arginine	Arg	R	225	−0.42
Asparagine	Asn	N	160	−0.28
Aspartic acid	Asp	D	150	−0.35
Cysteine	Cys	C	135	0.01
Glutamic acid	Glu	E	190	−0.25
Glutamine	Gln	Q	180	−0.21
Glycine	Gly	G	75	0.00
Histidine	His	H	195	−0.17
Isoleucine	Ile	I	175	0.15
Leucine	Leu	L	170	0.15
Lysine	Lys	K	200	−0.21
Methionine	Met	M	185	0.04
Phenylalanine	Phe	F	210	0.06
Proline	Pro	P	145	
Serine	Ser	S	115	−0.21
Threonine	Thr	T	140	−0.15
Tryptophan	Trp	W	255	0.03
Tyrosine	Tyr	Y	230	−0.03
Valine	Val	V	155	0.13

[a] Estimated with the rolling ball method.
[b] Hydrophilicity at 25°C is relative to glycine, and is based on the partitioning of a sidechain analogue between the two states (1 kJ g^{-1} = 0.24 kcal g^{-1}). The Gibbs free energy of transfer is given by $\Delta G_{\text{cyclohexane}\to\text{water}} = -RT\ln(c_{\text{water}}/c_{\text{cyclohexane}})$, where R is the universal gas constant, T is the absolute temperature, and c_{water} and $c_{\text{cyclohexane}}$ are the molar concentrations of sidechain analogues in the different phases (Radzicka and Wolfenden, 1988; Creighton, 1993; Privalov and Makhatadze, 1993).

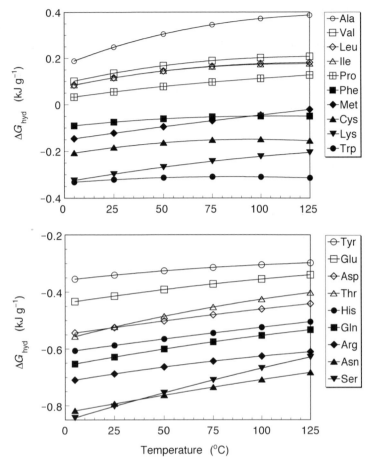

Fig. 3.4. Temperature dependence of the Gibbs free energy of transfer from vapor into water (hydrophilicity; $1 \text{ kJ g}^{-1} = 0.24 \text{ kcal g}^{-1}$) for uncharged (neutralized) amino acid sidechains (Privalov and Makhatadze, 1993).

The absence of hydrogen bonding between water and non-polar groups rather than the presence of favorable interactions between the non-polar groups themselves constitutes an important source of the protein stability in aqueous solution, the so-called hydrophobic interaction (Table 3.3; Figs. 3.4–3.6; Rose, 1987; Weber, 1996). Hydrophobicity and hydrophilicity usually are expressed as the Gibbs free energies of transfer from water into the reference state, and from a reference state into water, respectively. The transfer of the sidechains of hydrophobic amino acid residues, for example, leucine, isoleucine, and valine, from cyclohexane into water is energetically costly, and thus, the burial of hydrophobic sidechains in the folding reaction is energetically favorable. In contrast, hydrophilic sidechains, for example, that of arginine, prefer an aqueous environment over a hydrophobic environment, and are preferentially found at the surface in folded proteins.

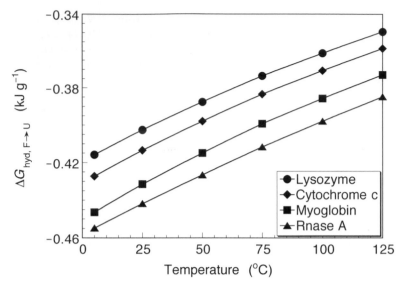

Fig. 3.5. Calculated temperature dependence of the change of the Gibbs free energy of hydration (hydrophilicity; 1 kJ g^{-1} = 0.24 kcal g^{-1}) of internal groups upon protein unfolding for horse heart cytochrome c, hen egg-white lysozyme, pancreatic ribonuclease A, and sperm-whale myoglobin, as indicated (Privalov and Makhatadze, 1993). $\Delta G_{\text{hyd,F}\rightarrow\text{U}}$ is negative because it is largely dominated by the contributions of polar groups (Privalov and Makhatadze, 1993).

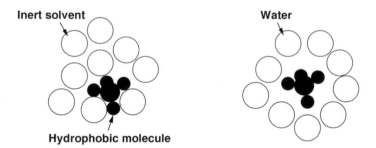

Fig. 3.6. The formation of cages around non-polar molecules in aqueous solution at low temperatures is connected with a decrease of entropy.

The Gibbs free energy of transfer, ΔG_{hyd}, of a non-polar molecule from a reference state, such as cyclohexane, into water (hydrophilicity) is composed of an enthalpy, ΔH_{hyd}, and entropy, $-T\Delta S_{\text{hyd}}$, term:

$$\Delta G_{\text{hyd}} = \Delta H_{\text{hyd}} - T\Delta S_{\text{hyd}} , \tag{3.5}$$

where T is the absolute temperature. At room temperature, ΔH_{hyd} for the transfer from cyclohexane into water is small and ΔG_{hyd} is dominated by the entropy term (Weber, 1996). This is mainly because the formation of ordered water cages around non-polar compounds is an entropically costly process, i.e., is connected with a decrease of entropy (Fig. 3.6).

Different chemical groups make vastly different contributions to the Gibbs free energies of transfer from organic solvent into water and of transfer from the gaseous phase into water (Fig. 3.4). For example, for non-cyclic structures, the Gibbs free energies of transfer from vapor into water for the groups $-CH_3$, $-CH_2-$, $-CH<$, $>C<$, $>C=O$, $-NH_2$, $-OH$, and $-NH-$ are 0.25, 0.05, -0.12, -0.41, -0.83, -1.48, -1.51, and -1.71 (all in kJ g^{-1}), respectively (Privalov and Makhatadze, 1993).

Below 100°C, the hydrophobic effect usually increases with temperature (Figs. 3.4, 3.5). At very high temperatures it does not further increase, but approaches a maximum, mainly because the structure-forming tendency of water, i.e., the entropic contribution to the hydrophobic effect, decreases with increasing temperature (Rose, 1987; Makhatadze and Privalov, 1993; Privalov and Makhatadze, 1993; Weber, 1996).

Intriguingly, studies on small organic compounds and proteins suggest that the change of Gibbs free energy of hydration of internal groups upon protein unfolding, $\Delta G_{hyd,F \to U}$, is negative for most proteins because $\Delta G_{hyd,F \to U}$ is largely dominated by the contributions of polar groups that prefer an aqueous over a hydrophobic environment (Fig. 3.5; Privalov and Makhatadze, 1993).

4 Calculation of the kinetic rate constants

Protein folding reactions can proceed according to a variety of different mechanisms. This chapter presents analytical solutions for kinetic rate constants and amplitudes for common reaction mechanisms.

The simplest case is that of a two-state transition, i.e., a reaction that proceeds without the occurrence of intermediates directly from the unfolded state, U, to the folded state, F (Sect. 4.2). In the transition region of the reaction $U \rightleftarrows F$, both forward and backward reaction contribute significantly to the observed rate constant (relaxation constant, decay constant). Under conditions that strongly favor folding (or unfolding), i.e., far outside the midpoint of equilibrium between folded and unfolded state, the transition can be treated as an irreversible reaction with the observed rate constant being dominated by the folding (or unfolding) rate constant.

For reversible three-state transitions, three cases have to be distinguished: 1. The intermediate, I, is on-pathway ($U \rightleftarrows I \rightleftarrows F$), i.e., is always passed through in the reaction from U to F (Sect. 4.3.1.1). 2. All species may interconvert, i.e., the transition from U to F may be passed through directly and also through the intermediate, I (Sect. 4.3.1.2). 3. I is off-pathway ($I \rightleftarrows U \rightleftarrows F$ or $U \rightleftarrows F \rightleftarrows I$), i.e., the reaction from U to F cannot proceed through I (Sect. 4.3.1.3).

Derivations of solutions for four-state transitions involve the treatment of cubic equations (Sect. 4.4).

Occasionally, folding reactions are linked with monomer–multimer transitions (Sect. 4.5). Examples are, (a) the protein is monomeric in the unfolded state but dimeric in the folded state, or (b) the protein aggregates in the unfolded, folded, or an intermediate state. Since these transitions affect the observed rate constants for folding events, solutions for a few simple cases are also presented.

Many important kinetic experiments (see Chaps. 5 and 10) involve the application of perturbation methods, such as small-amplitude temperature-jumping, repetitive pressure perturbation, ultrasonic velocimetry, and dielectric relaxation. These methods utilize a small perturbation of the chemical or physical equilibrium: A small change of physical or chemical conditions initiates a relaxation process to a new equilibrium. Since the amplitude of the perturbation is small, the mathematical treatment is tremendously simplified (Sect. 4.6).

The mathematical methods and analytical solutions presented for kinetic rate constants and amplitudes are not limited to protein folding reactions, but may be applied to a large variety of other chemical or physical reactions, for example, (a) in case of unimolecular mechanisms to conformational changes of other macromolecules (peptides, carbohydrates, lipids, DNA), and (b) in case of bimolecular mechanisms to aggregation-, enzyme–substrate binding-, and enzyme–inhibitor binding reactions.

Kinetic rate constants and amplitudes of unimolecular and bimolecular reactions are solutions of differential equations. Since no general mathematical formalism for the analytical solution of all differential equations has been found, the finding of a particular solution is often based on a mere guess that is confirmed by inserting it into the equation. For the confirmation of a solution as the general solution it is important to check whether it fulfills every possible initial condition.

Fortunately, the rate equations of unimolecular reactions are ordinary linear differential equations which generally have solutions that are linear combinations of exponential functions.

4.1
Transition state theory

The rate constant of the formation of a product, $k_{i \to f}$, in a step of the folding reaction (Fig. 4.1; Fersht, 1985; Matouschek et al., 1989) is, in good approximation,

$$k_{i \to f} = (k_B T / h) \exp(-\Delta G_{\#-i}/(RT)) , \qquad (4.1)$$

where $k_B = 1.3807 \times 10^{-23}$ J K^{-1} is the Boltzmann constant, $h = 6.6261 \times 10^{-34}$ J s is the Planck constant, T is the absolute temperature, $R = 8.3145$ J mol^{-1} K^{-1} is the molar gas constant, and $\Delta G_{\#-i}$ is the Gibbs free energy of activation.

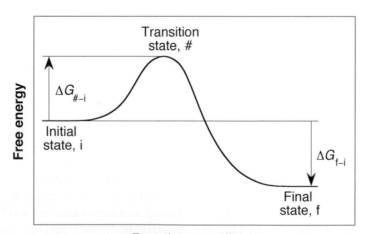

Reaction coordinate

Fig. 4.1. Transition state theory. The transition state is the state of highest energy along the reaction pathway that leads from the initial state (ground state) to the final state (product). The height of the transition state barrier determines the magnitude of the rate constant of transition ($\Delta G_{\#-i}$ and $\Delta G_{\#-f} = \Delta G_{\#-i} - \Delta G_{f-i}$ determine the rate constants of i→f and f→i, respectively).

The Gibbs free energy change of the reaction, ΔG_{f-i}, is connected with the equilibrium constant of the reaction, K_{f-i}, i.e., the ratio of product to reactant in equilibrium, by the well-known relation

$$\Delta G_{f-i} = \Delta H_{f-i} - T\Delta S_{f-i} = -RT \ln(K_{f-i}), \tag{4.2}$$

where ΔH_{f-i} is the enthalpy change and ΔS_{f-i} is the entropy change of the reaction.

4.2
Two-state transitions

4.2.1
Reversible two-state transition

To derive the rate equations for a reversible two-state transition between the states U and F

$$U \underset{k_{-1}}{\overset{k_1}{\rightleftarrows}} F, \tag{4.3}$$

we have to consider that the quantity of the decay of reactant, U, per time unit is proportional to the quantity of reactant itself and the quantity of the decay of product, F, per time unit is proportional to the quantity of product:

$$\frac{d[F]}{dt} = k_1[U] - k_{-1}[F] \tag{4.4}$$

$$\frac{d[U]}{dt} = k_{-1}[F] - k_1[U],$$

where $[U]$, $[F]$, k_1, k_{-1}, and t are the concentrations of U and F, the forward rate constant, the backward rate constant, and the time, respectively. Taking into account that the total concentration of species, $[UF] \equiv [U] + [F]$, is conserved, the rate equation for the change of the folded state may be written as

$$\frac{d[F]}{dt} = -(k_1 + k_{-1})[F] + k_1[UF] \tag{4.5}$$

$$[F](0) = [F_o],$$

where $[F_o]$ is the concentration of F at the start of the reaction, i.e., at $t = 0$. The solution of Eq. 4.5 is easily found by using the guess that the solution is a single-exponential function:

$$[F](t) = C_1 \exp(-k_1 t - k_{-1} t) + C_2 \tag{4.6}$$

$$C_1 = [F_o] - [UF]k_1/(k_1 + k_{-1})$$

$$C_2 = [UF]k_1/(k_1 + k_{-1})$$

$$[U](t) = [UF] - [F](t).$$

It can be shown that Eq. 4.6 fulfills every initial condition $[F_o] \in [0, [UF]]$ and represents the general solution.

Consequently, $[U](t)$ and $[F](t)$ follow single-exponential functions with an observed rate constant, $k_{obs} = k_1 + k_{-1}$ (Fig. 4.2, Table 4.1 in Sect. 4.7).

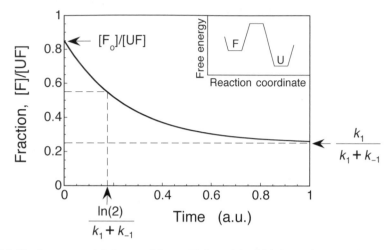

Fig. 4.2. Single-exponential change of the population of the folded state in a reversible two-state transition $(U \rightleftharpoons F)$ under conditions that favor unfolding. The parameters chosen for this example are: $k_1 = 1$, $k_{-1} = 3$, $[F_o]/[UF] = 0.85$. The observed rate constant (relaxation constant) is $k_{obs} = k_1 + k_{-1} = 4$. Inset: Energy landscape.

4.2.2
Irreversible two-state transition

Under conditions which strongly favor folding, the unfolding rate constant may be neglected:

$$U \xrightarrow{\ k_1\ } F \qquad (4.7)$$

The rate equation for F in this consecutive two-state transition is

$$\frac{d[F]}{dt} = k_1[U] \qquad (4.8)$$

$$[U] + [F] = [UF]$$

$$[F](0) = [F_o] \ ,$$

where $[U]$, $[F]$, k_1, and t are the concentrations of U and F, the forward rate constant, and the time, respectively. Here the solution is (see also Table 4.1)

$$[F](t) = ([F_o] - [UF]) \exp(-k_1 t) + [UF] \qquad (4.9)$$

$$[U](t) = ([UF] - [F_o]) \exp(-k_1 t) \ .$$

$[U](t)$ and $[F](t)$ follow single-exponential functions with an observed rate constant, $k_{obs} = k_1$. In contrast to reversible reactions, here the population of U vanishes with time (Fig. 4.3).

Analogously, for conditions that strongly favor unfolding, for example, at high concentrations of denaturant, we find $k_{obs} = k_{-1}$.

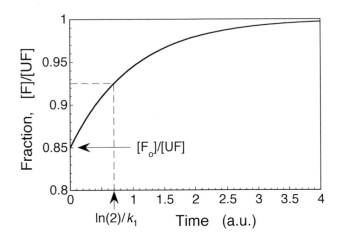

Fig. 4.3. Single exponential change of the fraction of the folded state, $[F]/[UF]$, in an irreversible two-state transition ($U \rightarrow F$). The parameters chosen for this example are: $k_1 = 1$; $[F_o]/[UF] = 0.85$.

4.3
Three-state transitions

4.3.1
Reversible three-state transitions

4.3.1.1
Reversible sequential three-state transition

For a reversible sequential three-state transition,

$$U \underset{k_{-1}}{\overset{k_1}{\rightleftharpoons}} I \underset{k_{-2}}{\overset{k_2}{\rightleftharpoons}} F \, , \tag{4.10}$$

between the states U, I, and F, with the positive rate constants, k_1, k_{-1}, k_2, and k_{-2}, the rate equations are

$$\frac{d[U]}{dt} = k_{-1}[I] - k_1[U] \tag{4.11}$$

$$\frac{d[F]}{dt} = k_2[I] - k_{-2}[F]$$

$$[I] = [UIF] - [U] - [F] \, ,$$

where [UIF] is the total concentration of the species U, I, and F, which is conserved in the reaction. Considering for simplicity first only the changes of [U], [I], and [F],

$$\frac{d\,\Delta[U]}{dt} = k_{-1}\Delta[I] - k_1\Delta[U] \tag{4.12}$$

$$\frac{d\,\Delta[F]}{dt} = k_2\Delta[I] - k_{-2}\Delta[F]$$

$$\Delta[I] = -\Delta[U] - \Delta[F] ,$$

and assuming that the solution of Eq. 4.12 is of the form

$$\Delta[U](t) = C_1 \exp(-\lambda t) \tag{4.13}$$

$$\Delta[F](t) = C_2 \exp(-\lambda t) ,$$

we obtain,

$$-\lambda\,\Delta[U] = -k_{-1}\,(\Delta[U] + \Delta[F]) - k_1\,\Delta[U] \tag{4.14}$$

$$-\lambda\,\Delta[F] = -k_2\,(\Delta[U] + \Delta[F]) - k_{-2}\,\Delta[F] .$$

By substituting $\Delta[U]$ or $\Delta[F]$ we find an equation for λ,

$$0 = \lambda^2 - \lambda\,(k_1 + k_{-1} + k_2 + k_{-2}) + k_1 k_2 + k_1 k_{-2} + k_{-1} k_{-2} , \tag{4.15}$$

which has two solutions,

$$\lambda_{1,2} = 0.5\,(k_1 + k_{-1} + k_2 + k_{-2}) \tag{4.16}$$

$$\pm\,((k_1 + k_{-1} + k_2 + k_{-2})^2 - 4\,(k_1 k_2 + k_1 k_{-2} + k_{-1} k_{-2}))^{1/2}) .$$

For physically permissible, i.e., positive rate constants, the term under the root cannot be negative. Thus, both solutions are real. Both $\lambda = \lambda_1$ and $\lambda = \lambda_2$ inserted into Eq. 4.13 fulfill Eq. 4.12, and thus, represent particular solutions. The general solution of Eq. 4.11 is a superposition of the two particular solutions:

$$[U](t) = C_1 \exp(-\lambda_1 t) + C_3 \exp(-\lambda_2 t) + C_5 \tag{4.17}$$

$$[F](t) = C_2 \exp(-\lambda_1 t) + C_4 \exp(-\lambda_2 t) + C_6 .$$

The constants, C_i, may be determined by inserting Eq. 4.17 into Eq. 4.11, and using the conservation relationship, $[UIF] = [U] + [I] + [F]$, and the initial conditions, $[U](0) = [U_o]$, $[F](0) = [F_o]$:

$$C_1 = ([F_o] - \xi_2[U_o] - C_6 + \xi_2 C_5) / (\xi_1 - \xi_2) \tag{4.18}$$

$$C_2 = \xi_1\,C_1$$

$$C_3 = ([F_o] - \xi_1[U_o] - C_6 + \xi_1 C_5) / (\xi_2 - \xi_1)$$

$$C_4 = \xi_2\,C_3$$

$$C_5 = [UIF]k_{-1}k_{-2}/(k_1 k_2 + k_1 k_{-2} + k_{-1} k_{-2})$$

$$C_6 = [UIF]k_1 k_2/(k_1 k_2 + k_1 k_{-2} + k_{-1} k_{-2})$$

$$\xi_1 = (\lambda_1 - k_1 - k_{-1}) / k_{-1}$$

$$\xi_2 = (\lambda_2 - k_1 - k_{-1}) / k_{-1} .$$

Alternatively, we can use the relations

$$C_1 = \xi_3 \, C_2 \qquad\qquad C_3 = \xi_4 \, C_4 \qquad\qquad (4.19)$$
$$\xi_3 = (\lambda_1 - k_2 - k_{-2}) \, / \, k_2 \qquad \xi_4 = (\lambda_2 - k_2 - k_{-2}) \, / \, k_2 \, .$$

From $1 = \xi_1 \xi_3$ follows $\xi_1 \neq 0$ and $\xi_3 \neq 0$, and thus, $C_2 = 0$ only if $C_1 = 0$, and $C_4 = 0$ only if $C_3 = 0$.

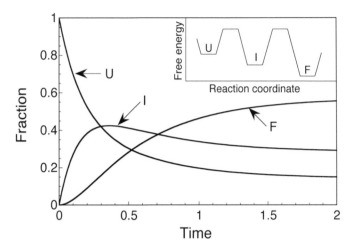

Fig. 4.4. Double-exponential change of the populations of folded, F, intermediate, I, and unfolded, U, state in a reversible sequential three-state transition (U $\rightleftarrows$ I $\rightleftarrows$ F) under conditions that favor folding. The parameters chosen for this example are: $k_1 = 4$, $k_{-1} = 2$, $k_2 = 2$, $k_{-2} = 1$, $[F_0]/[UIF] = 0$, $[U_0]/[UIF] = 1$. Observed rate constants ($\lambda \equiv k_{obs}$) are $\lambda_1 = 7$ and $\lambda_2 = 2$. Inset: Energy landscape.

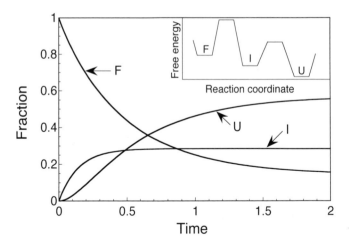

Fig. 4.5. Double-exponential change of the populations of folded, F, intermediate, I, and unfolded, U, state in a reversible sequential three-state transition (U $\rightleftarrows$ I $\rightleftarrows$ F) under conditions that favor unfolding. The parameters chosen for this example are: $k_1 = 2$, $k_{-1} = 4$, $k_2 = 1$, $k_{-2} = 2$, $[F_0]/[UIF] = 1$, $[U_0]/[UIF] = 0$. Observed rate constants ($\lambda \equiv k_{obs}$) are $\lambda_1 = 7$ and $\lambda_2 = 2$. Inset: Energy landscape.

Summarizing, the general solutions for the reversible sequential three-state transition (Eq. 4.10) usually are double-exponential functions, given by Eqs. 4.17 and 4.18 (see Table 4.1). Two special cases, where both [F] and [U] follow only single-exponential functions are: $C_1 = C_2 = 0$, and $C_3 = C_4 = 0$. In case $C_1 = C_2 = C_3 = C_4 = 0$, no kinetic event is macroscopically observed because the reaction is already in equilibrium.

In the example for Fig. 4.4, one can see that the intermediate accumulates kinetically. This is because the intermediate first becomes rapidly populated by the fast transition from U to I, but then becomes partially trapped by the high barrier of transition from I to F. In contrast, under conditions that favor unfolding, the early refolding intermediate is not accumulated, i.e., it is never populated above the level of population in equilibrium (Fig. 4.5).

After infinite time the system approaches an equilibrium,

$$[U](\infty) / [I](\infty) = K_{UI} = k_{-1} / k_1 \tag{4.20}$$
$$[I](\infty) / [F](\infty) = K_{IF} = k_{-2} / k_2 ,$$

where K_{UI}, K_{IF} are the equilibrium constants for unfolding. Using the conservation relation, $[I] = [UIF] - [U] - [F]$, we obtain the equilibrium concentrations:

$$[U](\infty) = [UIF] \, k_{-1}k_{-2} / (k_1k_2 + k_1k_{-2} + k_{-1}k_{-2}) \tag{4.21}$$
$$[I](\infty) = [UIF] \, k_1k_{-2} / (k_1k_2 + k_1k_{-2} + k_{-1}k_{-2})$$
$$[F](\infty) = [UIF] \, k_1k_2 / (k_1k_2 + k_1k_{-2} + k_{-1}k_{-2}) .$$

Rate constants for the special case of a much faster transition $U \rightleftarrows I$ than $I \rightleftarrows F$ are given in Sect. 8.3.3.2.

4.3.1.2
Reversible two-pathway three-state transition

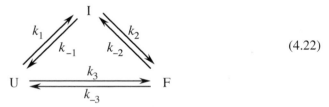

$$\tag{4.22}$$

The rate equations for the two-pathway three-state reaction (Eq. 4.22) are:

$$\frac{d[U]}{dt} = k_{-1}[I] - k_1[U] + k_{-3}[F] - k_3[U] \tag{4.23}$$

$$\frac{d[F]}{dt} = k_2[I] - k_{-2}[F] + k_3[U] - k_{-3}[F]$$

$$[I] = [UIF] - [U] - [F] .$$

The free energy difference between U and F must be independent of the pathway:

$$k_3/k_{-3} = k_1k_2/(k_{-1}k_{-2}) \tag{4.24}$$

The method of solving Eq. 4.23 is analogous to the method for the reversible sequential three-state transition (Sect. 4.3.1.1): Consider first for simplicity only the changes of [U], [I], and [F],

$$\frac{d\,\Delta[U]}{d\,t} = k_{-1}\Delta[I] - k_1\Delta[U] + k_{-3}\Delta[F] - k_3\Delta[U] \qquad (4.25)$$

$$\frac{d\,\Delta[F]}{d\,t} = k_2\Delta[I] - k_{-2}\Delta[F] + k_3\Delta[U] - k_{-3}\Delta[F]$$

$$\Delta[I] = -\Delta[U] - \Delta[F] \,,$$

and assume that a particular solution has the form:

$$\Delta[U](t) = C_1 \exp(-\lambda t) \qquad (4.26)$$

$$\Delta[F](t) = C_2 \exp(-\lambda t) \,.$$

This assumption leads to:

$$-\lambda\,\Delta[U] = -k_{-1}(\Delta[U] + \Delta[F]) - k_1\Delta[U] + k_{-3}\Delta[F] - k_3\Delta[U] \qquad (4.27)$$

$$-\lambda\,\Delta[F] = -k_2(\Delta[U] + \Delta[F]) - k_{-2}\Delta[F] + k_3\Delta[U] - k_{-3}\Delta[F] \,.$$

By simplifying Eq. 4.27 and substituting [U] or [F], we obtain an equation for λ,

$$0 = \lambda^2 - \lambda(k_1 + k_{-1} + k_2 + k_{-2} + k_3 + k_{-3}) + k_1k_2 + k_1k_{-2} \qquad (4.28)$$

$$+ k_1k_{-3} + k_{-1}k_{-2} + k_{-1}k_3 + k_{-1}k_{-3} + k_2k_3 + k_2k_{-3} + k_{-2}k_3 \,,$$

which has two real solutions (λ_1 corresponds to the positive sign before the root; the "$-$" corresponds to λ_2):

$$\lambda_{1,2} = 0.5\,(k_1 + k_{-1} + k_2 + k_{-2} + k_3 + k_{-3} \qquad (4.29)$$

$$\pm\,((k_1 + k_{-1} + k_2 + k_{-2} + k_3 + k_{-3})^2 - 4\,(k_1k_2 + k_1k_{-2} + k_1k_{-3} + k_{-1}k_{-2}$$

$$+ k_{-1}k_3 + k_{-1}k_{-3} + k_2k_3 + k_2k_{-3} + k_{-2}k_3))^{1/2})$$

$$[U](t) = C_1 \exp(-\lambda_1 t) + C_3 \exp(-\lambda_2 t) + C_5 \qquad (4.30)$$

$$[F](t) = C_2 \exp(-\lambda_1 t) + C_4 \exp(-\lambda_2 t) + C_6$$

$$C_1 = \begin{cases} ([F_o] - \xi_2[U_o] - C_6 + \xi_2 C_5)/(\xi_1 - \xi_2) & \text{for } k_{-1} \neq k_{-3} \\[2ex] [U_o] - [UIF]k_{-1}/(k_1 + k_{-1} + k_3) & \text{for } k_{-1} = k_{-3} \end{cases}$$

$$C_2 = \begin{cases} C_1(\lambda_1 - k_1 - k_{-1} - k_3)/(k_{-1} - k_{-3}) & \text{for } k_{-1} \neq k_{-3} \\[2ex] C_1(k_3 - k_2)/((k_2 + k_{-2})(k_1 - k_{-2})) & \text{for } k_{-1} = k_{-3}, k_1 \neq k_{-2} \\[2ex] 0 & \text{for } k_{-1} = k_{-3}, k_1 = k_{-2} \end{cases}$$

$$C_3 = \begin{cases} ([F_o] - \xi_1[U_o] - C_6 + \xi_1 C_5)/(\xi_2 - \xi_1) & \text{for } k_{-1} \neq k_{-3} \\[2ex] 0 & \text{for } k_{-1} = k_{-3} \end{cases}$$

$$C_4 = \begin{cases} C_3(\lambda_2 - k_1 - k_{-1} - k_3)/(k_{-1} - k_{-3}) & \text{for } k_{-1} \neq k_{-3} \\ [F_o] - C_2 - C_6 & \text{for } k_{-1} = k_{-3} \end{cases}$$

$$C_5 = [UIF](k_{-1}k_{-2} + k_{-1}k_{-3} + k_2 k_{-3}) / (k_1 k_2 + k_1 k_{-2} + k_1 k_{-3} + k_{-1}k_{-2} + k_{-1}k_3 + k_{-1}k_{-3} + k_2 k_3 + k_2 k_{-3} + k_{-2}k_3)$$

$$C_6 = [UIF](k_1 k_2 + k_{-1}k_3 + k_2 k_3) / (k_1 k_2 + k_1 k_{-2} + k_1 k_{-3} + k_{-1}k_{-2} + k_{-1}k_3 + k_{-1}k_{-3} + k_2 k_3 + k_2 k_{-3} + k_{-2}k_3)$$

$$\xi_1 = (\lambda_1 - k_1 - k_{-1} - k_3)/(k_{-1} - k_{-3}) \qquad \text{for } k_{-1} \neq k_{-3}$$
$$\xi_2 = (\lambda_2 - k_1 - k_{-1} - k_3)/(k_{-1} - k_{-3}) \qquad \text{for } k_{-1} \neq k_{-3}.$$

Similar to the sequential three-state transition (U$\rightleftarrows$I$\rightleftarrows$F), [U](t), [I](t), and [F](t) usually display a double-exponential behavior (Eq. 4.30) with two observed rate constants given by Eq. 4.29 (Figs. 4.6, 4.7, Table 4.1). An important implication is that the information of kinetic rate constants alone is generally not sufficient to distinguish between the two different mechanisms, two-pathway or sequential reaction. There are a few special cases in which [F](t) or [U](t) or both follow single-exponential functions. At the limits $k_3 \rightarrow 0$ and $k_{-3} \rightarrow 0$, Eqs. 4.29 and 4.30 transform into the solution for the reversible sequential three-state transition (Sect. 4.3.1.1).

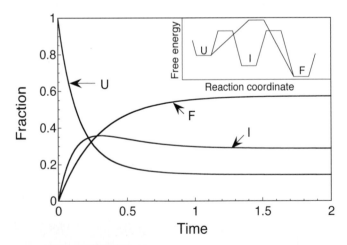

Fig. 4.6. Change of the populations of folded, F, intermediate, I, and unfolded state, U, in a reversible three-state transition with two parallel pathways (U$\rightleftarrows$I$\rightleftarrows$F and U$\rightleftarrows$F) under conditions that favor folding. The parameters chosen for this example are: $k_1 = 4$, $k_{-1} = 2$, $k_2 = 2$, $k_{-2} = 1$, $k_3 = 2$, $k_{-3} = 0.5$, $[F_o]/[UIF] = 0$, $[U_o]/[UIF] = 1$. Observed rate constants ($\lambda \equiv k_{obs}$) are $\lambda_1 = 8$ and $\lambda_2 = 3.5$. Inset: Energy landscape.

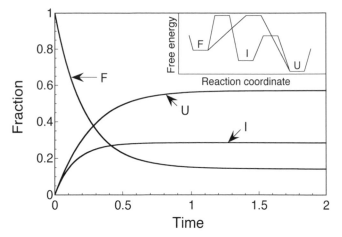

Fig. 4.7. Change of the populations of folded, F, intermediate, I, and unfolded state, U, in a reversible three-state transition with two parallel pathways ($U \rightleftarrows I \rightleftarrows F$ and $U \rightleftarrows F$) under conditions that favor unfolding. The parameters chosen for this example are: $k_1 = 2$, $k_{-1} = 4$, $k_2 = 1$, $k_{-2} = 2$, $k_3 = 0.5$, $k_{-3} = 2$, $[F_o]/[UIF] = 1$, $[U_o]/[UIF] = 0$. Inset: Energy landscape.

4.3.1.3
Reversible off-pathway intermediate

The third important type of reversible three-state transitions is illustrated in Eqs. 4.31 and 4.32. Strictly speaking, I is a side-product, but because it is often spectroscopically in-between U and F, it is commonly referred to as an off-pathway intermediate:

$$I \underset{k_{-1}}{\overset{k_1}{\rightleftarrows}} U \underset{k_{-2}}{\overset{k_2}{\rightleftarrows}} F \ , \qquad (4.31)$$

or

$$U \underset{k_{-1}}{\overset{k_1}{\rightleftarrows}} F \underset{k_{-2}}{\overset{k_2}{\rightleftarrows}} I \ . \qquad (4.32)$$

The solution for Eq. 4.31 is found by transforming it into Eq. 4.10 (Sect. 4.3.1.1) by exchanging I and U:

$$[I](t) = C_1 \exp(-\lambda_1 t) + C_3 \exp(-\lambda_2 t) + C_5 \qquad (4.33)$$

$$[F](t) = C_1 \xi_1 \exp(-\lambda_1 t) + C_3 \xi_2 \exp(-\lambda_2 t) + C_6$$

$$[U](t) = [UIF] - [I](t) - [F](t)$$

$$\lambda_{1,2} = 0.5 \, (k_1 + k_{-1} + k_2 + k_{-2} \qquad (4.34)$$

$$\pm ((k_1 + k_{-1} + k_2 + k_{-2})^2 - 4 \, (k_1 k_2 + k_1 k_{-2} + k_{-1} k_{-2}))^{1/2})$$

$$\xi_1 = (\lambda_1 - k_1 - k_{-1}) / k_{-1}$$

$$\xi_2 = (\lambda_2 - k_1 - k_{-1}) / k_{-1}$$
$$C_1 = ([F_o] - \xi_2[I_o] - C_6 + \xi_2 C_5) / (\xi_1 - \xi_2)$$
$$C_3 = ([F_o] - \xi_1[I_o] - C_6 + \xi_1 C_5) / (\xi_2 - \xi_1)$$
$$C_5 = [UIF]k_{-1}k_{-2}/(k_1 k_2 + k_1 k_{-2} + k_{-1}k_{-2})$$
$$C_6 = [UIF]k_1 k_2/(k_1 k_2 + k_1 k_{-2} + k_{-1}k_{-2}) \,,$$

where $[I_o] = [I](t{=}0)$, $[F_o] = [F](t{=}0)$, and $[UIF] = [U] + [I] + [F]$. Analogously, the solution for Eq. 4.32 is obtained by cyclically exchanging U, I, and F in Eq. 4.31. λ_1 and λ_2 are the same as for the mechanism $U \rightleftarrows I \rightleftarrows F$ (Eq. 4.16).

An important implication is that the information of kinetic rate constants alone is generally not sufficient to distinguish between on-pathway (Sect. 4.3.1.1) or off-pathway intermediates. A method for distinguishing these two cases is the Φ-value analysis (Sect. 8.3). Originally this method was designed for the structural resolution of transition states and intermediates (Goldenberg et al., 1989; Matouschek et al., 1989, 1990; Fersht et al., 1991, 1992; Matouschek and Fersht, 1991; Fersht, 1992, 1993, 1995a, b; Clarke and Fersht, 1993; Otzen et al., 1994; Itzhaki et al., 1995b, Nölting et al., 1995, 1997a; Nölting, 1998a, 1999), but it also provides information about the kinetic mechanism: In the Φ-value analysis mutants are used as reporters of structural consolidation along the folding pathway. In the case of an off-pathway intermediate, the calculated total Φ for the whole reaction from U to F is not 1 for all mutants when erroneously assuming an on-pathway mechanism (Sect. 8.3.4.2).

4.3.2
Irreversible three-state transitions

4.3.2.1
Irreversible consecutive three-state transition

$$U \xrightarrow{\;\;k_1\;\;} I \xrightarrow{\;\;k_2\;\;} F \qquad\qquad (4.35)$$

The solution (Fig. 4.8, Table 4.1) is derived by using the methods presented in the previous sections:

$$[U](t) = [U_o] \exp(-k_1 t) \qquad\qquad (4.36)$$

$$[F](t) = C_2 \exp(-k_1 t) + C_4 \exp(-k_2 t) + [UIF]$$

$$C_2 = \begin{cases} k_2[U_o]/(k_1 - k_2) & \text{for } k_1 \neq k_2 \\[2mm] -k_1[U_o]t & \text{for } k_1 = k_2 \end{cases}$$

$$C_4 = \begin{cases} [F_o] - k_2[U_o]/(k_1 - k_2) - [UIF] & \text{for } k_1 \neq k_2 \\[2mm] [F_o] - [UIF] & \text{for } k_1 = k_2 \end{cases}$$

$$[U_o] = [U](0)$$
$$[F_o] = [F](0)$$
$$[UIF] = [U] + [I] + [F] \, .$$

The definitions are as in Sect. 4.3.1. One should mention, that in case $k_1 = k_2$, C_2 is a function of time, and thus, $[F](t)$ and $[I](t)$ are not pure superpositions of two exponential functions anymore.

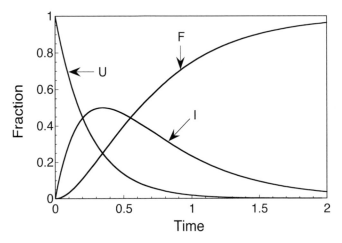

Fig. 4.8. Change of the populations of folded, F, intermediate, I, and unfolded, U, state in an irreversible three-state transition (U→I→F). The parameters chosen for this example are: $k_1 = 4$, $k_2 = 2$, $[F_o]/[UIF] = 0$, $[U_o]/[UIF] = 1$.

4.3.2.2
Irreversible parallel decay

$$F_1 \xleftarrow{\;k_1\;} U \xrightarrow{\;k_2\;} F_2 \qquad (4.37)$$

This type of reaction may occur under conditions that strongly favor folding when two conformations (e.g., a correctly folded and a misfolded, see Sect. 9.3) are produced. Here the populations of the species are

$$[U](t) = [U_o] \exp(-(k_1 + k_2)t) \qquad (4.38)$$
$$[F_1](t) = -[U_o] \exp(-(k_1 + k_2)t) \, k_1 \, / \, (k_1 + k_2) + C_1$$
$$[F_2](t) = -[U_o] \exp(-(k_1 + k_2)t) \, k_2 \, / \, (k_1 + k_2) + C_2 \, ,$$

where the definitions are analogous to those in Sect. 4.3.1, and the constants C_1 and C_2 depend on the initial conditions.

4.4
Reversible sequential four-state transition

$$U \underset{k_{-1}}{\overset{k_1}{\rightleftharpoons}} I_1 \underset{k_{-2}}{\overset{k_2}{\rightleftharpoons}} I_2 \underset{k_{-3}}{\overset{k_3}{\rightleftharpoons}} F \tag{4.39}$$

The solution of this case involves the treatment of a cubic equation (Eq. 4.40) (Beyer, 1991). With the exception of a few special cases, a superposition of three exponential functions is observed:

$$0 = \lambda^3 + p\lambda^2 + q\lambda + r \tag{4.40}$$

$$p = -(k_1 + k_{-1} + k_2 + k_{-2} + k_3 + k_{-3})$$

$$q = k_1 k_2 + k_1 k_{-2} + k_1 k_3 + k_1 k_{-3} + k_{-1} k_{-2} + k_{-1} k_3 + k_{-1} k_{-3} + k_2 k_3 + k_2 k_{-3} + k_{-2} k_{-3}$$

$$r = -(k_1 k_2 k_3 + k_1 k_2 k_{-3} + k_1 k_{-2} k_{-3} + k_{-1} k_{-2} k_{-3})$$

$$\lambda_1 = A + B - p/3$$

$$\lambda_{2,3} = -0.5(A + B) \pm 0.5(A - B)\sqrt{-3} - p/3$$

$$A = \sqrt[3]{-0.5b + c} \qquad B = \sqrt[3]{-0.5b - c} \qquad c = \sqrt{\frac{b^2}{4} + \frac{a^3}{27}}$$

$$a = (3q - p^2)/3 \qquad b = (2p^3 - 9pq + 27r)/27$$

$$[U](t) = C_1 \exp(-\lambda_1 t) + C_3 \exp(-\lambda_2 t) + C_5 \exp(-\lambda_3 t) + C_7$$

$$[F](t) = C_2 \exp(-\lambda_1 t) + C_4 \exp(-\lambda_2 t) + C_6 \exp(-\lambda_3 t) + C_8 .$$

For example, for $k_1 = 4$, $k_{-1} = 2$, $k_2 = 2$, $k_{-2} = 1$, $k_3 = 2$, $k_{-3} = 0.5$, one obtains $c = -5.449i$, $A = 1.630 - 0.733i$, $B = 1.630 + 0.733i$, where the imaginary number i is defined as $i \equiv \sqrt{-1}$, and $\lambda_1 = 7.093$, $\lambda_2 = 3.473$, $\lambda_3 = 0.934$. Compared with the example for three-state transitions with two parallel pathways (Fig. 4.6), the observed rate constants, λ_1 and λ_2, are slower, and there is an additional, even slower, phase, λ_3.

Concerning the calculation of the roots in Eq. 4.40, it should be mentioned that a complex number $z = z_r + iz_i$ can be transformed to $z = r \times (\cos \phi + i \sin \phi)$, where $r = (z_r^2 + z_i^2)^{0.5}$ and $\phi = \arctan(z_i/z_r)$. From *de Moivre's* identity, $(e^{i\phi})^n = e^{in\phi}$, and *Euler's* formula, $e^{i\phi} = \cos \phi + i \sin \phi$, follows $(\cos \phi + i \sin \phi)^n = \cos(n\phi) + i \sin(n\phi)$. Thus,

$$z^{1/n} = (r \times (\cos \phi + i \sin \phi))^{1/n} \tag{4.41}$$

$$= r^{1/n} \times (\cos((\phi + 2\pi k)/n) + i \sin((\phi + 2\pi k)/n))$$

$$n = 2, 3, \ldots ; \qquad k = 0, 1, \ldots , n-1$$

$$z = z_r + iz_i \qquad r = (z_r^2 + z_i^2)^{0.5} \qquad \phi = \arctan(z_i/z_r) .$$

For example, the three third roots of 1 are 1 and $-1/2 \pm \sqrt{-3}/2$.

4.5
Reactions with monomer–dimer transitions

4.5.1
Monomer–dimer transition

$$2A \underset{k_{-1}}{\overset{k_1}{\rightleftharpoons}} A_2 \tag{4.42}$$

In contrast to the cases treated in the previous sections, here we are encountering a bimolecular reaction. The rate equation for [A] is a first-order non-linear differential equation:

$$\frac{d[A]}{dt} = -2k_1[A]^2 - k_{-1}[A] + k_{-1}[A_{tot}] \tag{4.43}$$

$$[A](0) = [A_o]$$

$$[A_2] = 0.5 \, ([A_{tot}] - [A]) \, ,$$

where [A$_{tot}$] is the total protein concentration in equivalents of monomers, and the other definitions are analogous to those in Sect. 4.2. We start with the guess that particular solutions are of the types:

$$[A] = B + C \tanh[D(t + E)] \tag{4.44}$$

$$[A] = B + C \coth[D(t + E)]$$

$$\tanh(u) \equiv \frac{\exp(u) - \exp(-u)}{\exp(u) + \exp(-u)}$$

$$\coth(u) \equiv \frac{\exp(u) + \exp(-u)}{\exp(u) - \exp(-u)} \, .$$

By inserting Eq. 4.44 into Eq. 4.43 and using the relations:

$$d[\tanh(u)]/dt = (1 - \tanh^2(u)) \times du/dt \tag{4.45}$$

$$d[\coth(u)]/dt = (1 - \coth^2(u)) \times du/dt \, ,$$

the guess is verified and the constants are calculated (Figs. 4.9, 4.10):

$$B = -0.25 \, [A_{tot}] \, K \tag{4.46}$$

$$C = 0.25 \, [A_{tot}] \, (K^2 + 8K)^{0.5}$$

$$D = 2 \, k_1 \, C$$

$$K = k_{-1}/([A_{tot}]k_1)$$

$$E = \begin{cases} \text{arctanh}[([A_o] - B)/C]/D & \text{for } \left| ([A_o] - B)/C \right| < 1 \\ \\ \text{arccoth}[([A_o] - B)/C]/D & \text{for } \left| ([A_o] - B)/C \right| > 1 \end{cases}$$

$$[A] = \begin{cases} B + C \tanh[D(t + E)] & \text{for } |([A_o] - B)/C| < 1 \\ B + C \coth[D(t + E)] & \text{for } |([A_o] - B)/C| > 1 . \end{cases}$$

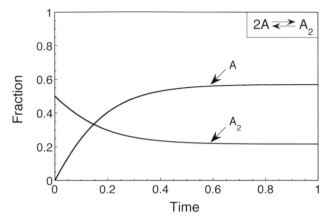

Fig. 4.9. Change of the populations of monomers and dimers in a monomer–dimer transition $(2A \rightleftarrows A_2)$. The parameters chosen for this example are: $k_1 = 2/[A_{tot}]$, $k_{-1} = 3$, $[A_o]/[A_{tot}] = 0$.

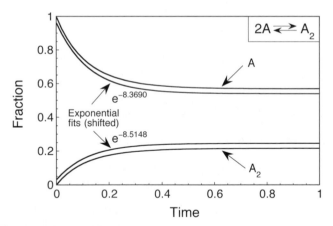

Fig. 4.10. Change of the populations of monomers and dimers in a monomer–dimer transition $(2A \rightleftarrows A_2)$, compared with single-exponential fits which are shifted by 0.03 units for better visibility. Only little differences are observed between the kinetic traces and the wrong curve fits. Thus, the information about the shape of the kinetic traces is often not sufficient to distinguish between bimolecular and unimolecular reactions. The parameters chosen for this example are: $k_1 = 2/[A_{tot}]$, $k_{-1} = 3$, $[A_o]/[A_{tot}] = 1$.

Alternatively, the solution of Eq. 4.43 may be represented as a linear combination of both particular solutions (Eq. 4.44). The equilibrium concentrations are:

$$[A](\infty) = 0.25 \, [A_{tot}] \, ((K^2 + 8K)^{0.5} - K) \qquad (4.47)$$
$$[A_2](\infty) = 0.5 \, ([A_{tot}] - [A](\infty)) \, .$$

In contrast to the first-order reactions in Sects. 4.2–4.4, the rate constants for reactions that involve a monomer–dimer transition are always dependent on the concentration.

4.5.2
Reversible two-state folding transition linked with a monomer–dimer transition

Another common case of a kinetic reaction is a reversible two-state folding transition of monomeric molecules followed by the dimerization in the folded state (Fig. 4.11):

$$2U \underset{k_{-1}}{\overset{k_1}{\rightleftharpoons}} 2F \underset{k_{-2}}{\overset{k_2}{\rightleftharpoons}} F_2 \qquad (4.48)$$

The rate equations are:

$$\frac{d[U]}{dt} = k_{-1}[F] - k_1[U] \qquad (4.49)$$

$$\frac{d[F_2]}{dt} = k_2 [F]^2 - k_{-2}[F_2]$$
$$[F] = [UF] - [U] - 2[F_2] \, ,$$

where [UF] is the total concentration in equivalents of monomers. The rate equation for [U] is quite complicated:

$$k_{-1}\frac{d^2[U]}{dt^2} + 2k_2\left(\frac{d[U]}{dt}\right)^2 + 4k_1k_2[U]\frac{d[U]}{dt} + k_{-1}(k_1 + k_{-1} + k_{-2})\frac{d[U]}{dt} \qquad (4.50)$$
$$+2k_1^2k_2[U]^2 + k_{-1}k_{-2}(k_1 + k_{-1})[U] - k_{-1}^2 k_{-2}[UF] = 0$$

Exponential functions are solutions only for a few special cases. A very crude approximation for [U] which is valid for several non-trivial cases is given by:

$$[U] = C + \frac{\sum\limits_{i=1}^{n} A_i \, e^{-i\lambda t}}{\sum\limits_{i=1}^{n} B_i \, e^{-i\lambda t}} \, , \qquad (4.51)$$

where $n \geq 4$, A_i, B_i, and C are constants. By inserting Eq. 4.51 into Eq. 4.50 and using the boundary condition for [U](0), one obtains several algebraic equations for A_i, B_i, and C. These can be solved for the time-independent terms and the terms which contain only low-order exponential functions. Then the error due to the non-equality of the terms which contain high-order exponential functions is occasionally found to be relatively small.

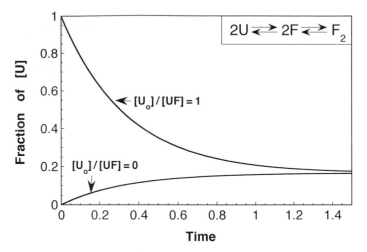

Fig. 4.11. Change of the population of the unfolded state in a two-state folding transition that is in sequence with a monomer–dimer transition ($2U \rightleftharpoons 2F \rightleftharpoons F_2$). The parameters chosen for this example are: $k_1 = 3$, $k_{-1} = 1$, $k_2 = 2/[UF]$, $k_{-2} = 3$. Two cases are shown: $[U_o]/[UF] = 1$, and $[U_o]/[UF] = 0$.

4.6
Kinetic rate constants for perturbation methods

The mathematical treatment of bimolecular reaction kinetics is tremendously simplified for small-amplitude perturbation methods, such as small-amplitude temperature-jumping, pressure perturbation, ultrasonic velocimetry, and dielectric relaxation (Table 4.2). In these methods, the changes of the populations of the states are small because the equilibrium of the species involved is only slightly perturbed by a small change in the physical or chemical conditions. Consider, for example, Eq. 4.52:

$$2U \; \underset{k_{-1}}{\overset{k_1}{\rightleftharpoons}} \; 2F \; \underset{k_{-2}}{\overset{k_2}{\rightleftharpoons}} \; F_2 \tag{4.52}$$

In equilibrium, per definition, the macroscopic changes in concentration are zero:

$$\frac{d[U]}{dt} = k_{-1}[F] - k_1[U] = 0 \tag{4.53}$$

$$\frac{d[F_2]}{dt} = k_2[F]^2 - k_{-2}[F_2] = 0$$

$$[F] = [UF] - [U] - 2[F_2] \,,$$

where [UF] is the total concentration in equivalents of monomers. Thus, the equilibrium concentrations are:

$$[U]_{eq} = \frac{k_{-1}}{4k_1^2 k_2}(\sqrt{(k_1 + k_{-1})^2 k_{-2}^2 + 8k_1^2 k_2 k_{-2}[UF]} - (k_1 + k_{-1})k_{-2}) \quad (4.54)$$

$$[F]_{eq} = \frac{1}{4k_1 k_2}(\sqrt{(k_1 + k_{-1})^2 k_{-2}^2 + 8k_1^2 k_2 k_{-2}[UF]} - (k_1 + k_{-1})k_{-2})$$

$$[F_2]_{eq} = 0.5([UF] - [U]_{eq} - [F]_{eq}) .$$

Upon perturbation, the equilibrium moves to a new position:

$$[U] = [U]_{eq} + [u] \quad (4.55)$$

$$[F] = [F]_{eq} + [f] \qquad [F_2] = [F_2]_{eq} + [f_2] ,$$

where are [u], [f], and $[f_2]$ are the small changes of $[U]_{eq}$, $[F]_{eq}$, and $[F_2]_{eq}$, respectively. Thus, the relaxation of $[F_2]$ is described by

$$\frac{d[F_2]_{eq}}{dt} + \frac{d[f_2]}{dt} = k_2([F]_{eq} + [f])^2 - k_{-2}([F_2]_{eq} + [f_2]) , \quad (4.56)$$

and so,

$$\frac{d[f_2]}{dt} = k_2(2[f][F]_{eq} + [f]^2) - k_{-2}[f_2] . \quad (4.57)$$

Neglecting the second-order term, $[f]^2$, leads to

$$\frac{d[f_2]}{dt} = 2k_2[f][F]_{eq} - k_{-2}[f_2] . \quad (4.58)$$

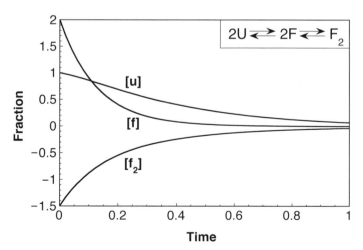

Fig. 4.12. Change in the populations of the monomeric unfolded state, monomeric folded state, and dimeric folded state upon a small perturbation of their equilibrium ($2U \rightleftarrows 2F \rightleftarrows F_2$). [u], [f], and $[f_2]$ are the small changes in the equilibrium concentrations, $[U]_{eq}$, $[F]_{eq}$, and $[F_2]_{eq}$, for the states U, F, and F_2, respectively. Parameters chosen for this example are: $k_1 = 3$, $k_{-1} = 1$, $k_2 = 2/[UF]$, $k_{-2} = 3$, $2[u_o] = [f_o]$. Observed rate constants ($\lambda \equiv k_{obs}$) are $\lambda_1 = 8$ and $\lambda_2 = 3$.

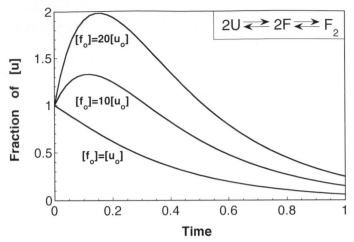

Fig. 4.13. Change in the population of the unfolded state upon a small perturbation of the equilibrium between monomeric unfolded, monomeric folded, and dimeric folded state ($2U \rightleftarrows 2F \rightleftarrows F_2$). The parameters chosen for this example are: $k_1 = 3$, $k_{-1} = 1$, $k_2 = 2/[UF]$, $k_{-2} = 3$, and $[f_o] = [u_o]$, $[f_o] = 10 [u_o]$, $[f_o] = 20 [u_o]$, respectively, as indicated. $[u_o]$ and $[f_o]$ depend on the magnitude and type of the perturbation and on the physical and chemical properties of the involves species. Observed rate constants ($\lambda \equiv k_{obs}$) are $\lambda_1 = 8$ and $\lambda_2 = 3$.

Analogously,

$$\frac{d[u]}{dt} = k_{-1}[f] - k_1[u] \tag{4.59}$$

$$[f] + [u] + 2[f_2] = 0 .$$

By combination of Eqs. 4.58 and 4.59 we obtain:

$$\frac{d^2[u]}{dt^2} + \frac{d[u]}{dt}(k_1 + k_{-1} + k_{-2} + 4k_2[F]_{eq}) \tag{4.60}$$
$$+ [u](k_1 k_{-2} + k_{-1}k_{-2} + 4k_1 k_2[F]_{eq}) = 0$$

$$[u](t) = [u_o]C \exp(-\lambda_1 t) + [u_o](1 - C) \exp(-\lambda_2 t) \tag{4.61}$$
$$C = (k_{-1}[f_o]/[u_o] + \lambda_2 - k_1)/(\lambda_2 - \lambda_1)$$
$$\lambda_{1,2} = 0.5p \pm (0.25p^2 - q)^{0.5}$$
$$p = k_1 + k_{-1} + k_{-2} + 4k_2[F]_{eq}$$
$$q = k_1 k_{-2} + k_{-1}k_{-2} + 4k_1 k_2[F]_{eq}$$

$$[f] = (\frac{d[u]}{dt} + k_1[u])/k_{-1}$$
$$= ([u_o]/k_{-1})\{C(k_1 - \lambda_1)\exp(-\lambda_1 t) + (1 - C)(k_1 - \lambda_2)\exp(-\lambda_2 t)\}$$

$$[f_2] = -0.5([f] + [u]) \, ,$$

where $[u_o]$ and $[f_o]$ change with magnitude and type of the perturbation, and depend on the physical and chemical properties of the species involved. Depending on these parameters, convex or concave curve shapes, with or without transient accumulation, are observed (Figs. 4.12, 4.13).

Analogously, the simple case,

$$2F \; \underset{k_{-1}}{\overset{k_1}{\rightleftharpoons}} \; F_2 \; , \tag{4.62}$$

is easily derived:

$$\frac{d[f_2]}{dt} = -4k_1[f_2][F]_{eq} - k_{-1}[f_2] \tag{4.63}$$

$$[F]_{eq} = 0.25(\sqrt{K^2 + 8K[F_{tot}]} - K)$$

$$[F_{tot}] = [F] + 2[F_2]$$

$$K = k_{-1} / k_1 \; .$$

Here the observed rate constant, λ, is

$$\lambda = k_{-1} + 4k_1[F]_{eq} \; . \tag{4.64}$$

The rate constants of unimolecular reactions are the same as those derived in Sects. 4.2–4.4 (see Table 4.1).

4.7
Summary

In reactions which involve solely unimolecular transitions, so-called first-order reactions, the fractions of the species involved usually change with time according to superpositions of exponential functions with concentration-independent rate constants (Table 4.1).

As soon as bimolecular reactions are at least partly involved, the speed of reaction becomes concentration-dependent (Table 4.2). However, under pseudo-first-order experimental conditions this concentration-dependence may be undetectably small, for example, under conditions where the process is dominated by a dissociation event.

The kinetic traces of bimolecular reactions often have non-exponential shapes. However, in small perturbation methods, generally, exponential shapes of the kinetic traces are observed also for bimolecular reactions.

Table 4.1. Rate constants for first-order reaction mechanisms. For special cases and for magnitudes of changes of reactants and products see Sects. 4.2–4.4.

Reaction	Observed rate constants, $\lambda \equiv k_{obs}$[a] (relaxation constants)
$U \xrightarrow{k_1} F$	$\lambda = k_1$
$U \xrightleftharpoons[k_{-1}]{k_1} F$	$\lambda = k_1 + k_{-1}$
$U \xrightarrow{k_1} I \xrightarrow{k_2} F$	$\lambda_{1,2} = k_1, k_2$
$U \xrightleftharpoons[k_{-1}]{k_1} I \xrightleftharpoons[k_{-2}]{k_2} F$	$\lambda_{1,2} = 0.5 \, (k_1 + k_{-1} + k_2 + k_{-2})$ $\pm \, ((k_1 + k_{-1} + k_2 + k_{-2})^2$ $- \, 4 \, (k_1 k_2 + k_1 k_{-2} + k_{-1} k_{-2}))^{1/2})$

For the triangular mechanism (U, I, F with $k_1, k_{-1}, k_2, k_{-2}, k_3, k_{-3}$):

$$\lambda_{1,2} = 0.5 \, (k_1 + k_{-1} + k_2 + k_{-2} + k_3 + k_{-3})$$
$$\pm \, ((k_1 + k_{-1} + k_2 + k_{-2} + k_3 + k_{-3})^2$$
$$- \, 4 \, (k_1 k_2 + k_1 k_{-2} + k_1 k_{-3} +$$
$$k_{-1} k_{-2} + k_{-1} k_3 + k_{-1} k_{-3} + k_2 k_3$$
$$+ \, k_2 k_{-3} + k_{-2} k_3))^{1/2})$$

For the mechanism $U \xrightleftharpoons[k_{-1}]{k_1} I_1 \xrightleftharpoons[k_{-2}]{k_2} I_2 \xrightleftharpoons[k_{-3}]{k_3} F$:

$$\lambda_1 = A + B - p/3$$
$$\lambda_{2,3} = -\,0.5(A + B)$$
$$\pm \, 0.5(A - B)\sqrt{-3} - p/3$$
$$A = \sqrt[3]{-0.5b + c} \quad B = \sqrt[3]{-0.5b - c}$$
$$c = \sqrt{\frac{b^2}{4} + \frac{a^3}{27}}$$
$$a = (3q - p^2)/3$$
$$b = (2p^3 - 9pq + 27r)/27$$
$$p = -(k_1 + k_{-1} + k_2 + k_{-2} + k_3 + k_{-3})$$
$$q = k_1 k_2 + k_1 k_{-2} + k_1 k_3 + k_1 k_{-3} + k_{-1} k_{-2}$$
$$+ \, k_{-1} k_3 + k_{-1} k_{-3} + k_2 k_3 + k_2 k_{-3} + k_{-2} k_{-3}$$
$$r = -(k_1 k_2 k_3 + k_1 k_2 k_{-3}$$
$$+ \, k_1 k_{-2} k_{-3} + k_{-1} k_{-2} k_{-3})$$

[a] In special cases, some of the given rate constants do not apply.

Table 4.2. Rate constants for second-order reaction mechanisms in small-amplitude perturbation methods (see Sect. 4.6).

Reaction	Observed rate constants, $\lambda \equiv k_{obs}{}^a$ (relaxation constants)
$2F \underset{k_{-1}}{\overset{k_1}{\rightleftarrows}} F_2$	$\lambda = k_{-1} + 4k_1[F]_{eq}$
$A + B \underset{k_{-1}}{\overset{k_1}{\rightleftarrows}} AB$	$\lambda = k_{-1} + k_1([A]_{eq} + [B]_{eq})$
$2U \underset{k_{-1}}{\overset{k_1}{\rightleftarrows}} 2F \underset{k_{-2}}{\overset{k_2}{\rightleftarrows}} F_2$	$\lambda_{1,2} = 0.5p \pm (0.25p^2 - q)^{0.5}$ $p = k_1 + k_{-1} + k_{-2} + 4k_2[F]_{eq}$ $q = k_1k_{-2} + k_{-1}k_{-2} + 4k_1k_2[F]_{eq}$

^{a} The rate constants are calculated for experimental conditions under which a chemical equilibrium is approached prior to application of a small perturbation. $[F]_{eq}$, $[A]_{eq}$, and $[B]_{eq}$ are the equilibrium concentrations of F, A, and B, respectively. In special cases, some of the given rate constants do not apply.

5 High kinetic resolution of protein folding events

5.1
Ultrafast mixing

One of the oldest ways of inducing rapid protein folding is to mix a solution of unfolded protein with buffers that favor folding (Fig. 5.1). The reaction is followed by an optical probe, for example, ultraviolet–visible or infrared absorption, circular dichroism (CD), scattering, or fluorescence. In particular, CD detection in combination with rapid mixing is an exquisitely sensitive probe of conformational changes (Luchins and Beychok, 1978; Pflumm et al., 1986; Kuwajima et al., 1987, 1993, 1996; Elöve et al., 1992; Kalnin and Kuwajima, 1995; Arai and Kuwajima, 1996). Kinetic resolution of molecular dimensions became possible by advances in X-ray scattering (Semisotnov et al., 1996) and dynamic light scattering (Gast et al., 1997). H/D exchange kinetics is frequently followed by nuclear magnetic resonance (NMR) spectrometry to obtain information about local and global folding events (see Sects. 7.1 and 8.1). Mass spectrometry detection of H/D exchange requires significantly smaller quantities of protein (Miranker et al., 1996a, b). Real-time NMR spectroscopy with kinetic resolution has significantly advanced into the millisecond time range (Frieden et al., 1993; Balbach et al., 1995; Hoeltzli and Frieden, 1995).

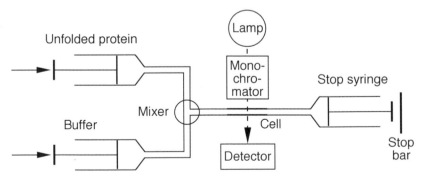

Fig. 5.1. Stopped-flow method. Two syringes containing a solution of unfolded protein and a buffer that favors refolding, respectively, are simultaneously pushed. Turbulence in the T-mixer, where both liquids join together, causes a rapid mixing. After a certain amount of liquid has passed through the mixer, the position of the stop syringe triggers a stop signal and a signal for the detector to start recording the reaction kinetics in the sample cell.

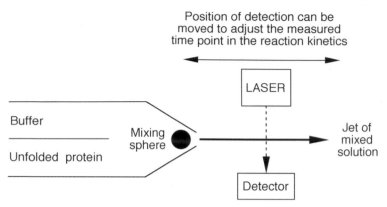

Position of detection can be
moved to adjust the measured
time point in the reaction kinetics

Fig. 5.2. Ultrafast continuous-flow mixing. A solution of unfolded protein and a buffer that favors folding are gently joined together and passed through a tube with a decreasing cross-section. At the end of the tube, the laminar flow is changed into a highly turbulent flow by passing the liquid over a sphere of only a few 10 μm diameter. The mixed solution forms a continuous free jet. Each position in the jet corresponds to a certain time point in the reaction kinetics. Kinetic traces are recorded by moving the LASER/detector system along the jet. At a typical flow speed of 10–100 m s^{-1}, a change of the position by 100 μm corresponds to a change in time of 1–10 μs. Because in free air the jet is stable over a few cm, reactions may be followed from microseconds to milliseconds (Regenfuss et al., 1985; Regenfuss and Clegg, 1987).

Unfortunately, the dead time of common stopped-flow devices is usually around 1 millisecond. Faster mixing requires stronger turbulence in the mixing chamber. Decreasing the size of the tubes and increasing the speed of flow would require impracticably high pressures in the common design.

In Thomas Jovin's group (Regenfuss et al., 1985; Regenfuss and Clegg, 1987) it has been realized that mixing in microseconds can be performed by using a number of innovations (Fig. 5.2 and Table 5.1 in Sect. 5.8): 1. In order to reduce the flow resistance, relatively wide tubes are used for the transport of the two liquids to the mixing chamber. Before reaching the mixing chamber, the two streams of liquid are gently joined together so that no significant mixing occurs. 2. The mixing chamber consists of a tip in which a sphere is placed. Intense mixing is triggered by the liquid flowing over the sphere. 3. The size of the mixing sphere may be as small as a few micrometers. At a typical flow speed of 10–100 m s^{-1}, a dead length of 100 μm corresponds to a dead time of 1–10 μs. 4. Using continuous-flow rather than stopped-flow avoids pressure waves at high flow speeds.

These modifications led to a 100-fold reduction of the mixing time down to 10 μs. According to theoretical considerations, further improvement to 1 μs should be possible by decreasing the size of the mixer (Regenfuss et al., 1985).

Rousseau et al. (Takahashi et al., 1997; Yeh et al., 1997; Yeh and Rousseau, 1998) modified the classic design (Fig. 5.1) by choosing a narrow cross-section for the mixer, 100 μm × 25 μm, compared with 250 μm × 250 μm for the

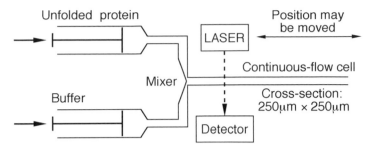

Fig. 5.3. Ultrafast mixing using a continuous-flow cell. In order to reduce the flow resistance, relatively wide tubes of several 100 μm inner diameter connect the syringes with the T-mixer. At the position of the mixer, the cross-section of the tubes is decreased down to 100 μm × 25 μm. Different positions of the LASER/detector system correspond to different time points of the reaction kinetics triggered by the mixing of the two liquids. Using a continuous-flow cell instead of a free jet in air (see Fig. 5.2) improves the optical stability but slightly increases the necessary pressure (Takahashi et al., 1997; Yeh et al., 1997; Yeh and Rousseau, 1998). See also Lin et al. (2003); Cherepanov and De Vries (2004).

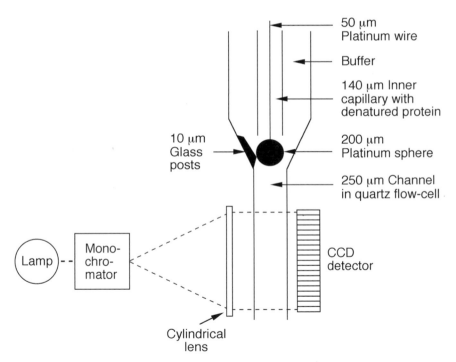

Fig. 5.4. Ultrafast continuous-flow mixing head (Shastry et al., 1988; Sauder et al., 1996; Park et al., 1997; Shastry and Roder, 1998). Mixing over a sphere of 200 μm diameter enhances turbulence and enables mixing in the 15-μs time scale and a dead time of about 50 μs. A 50-μm platinum wire serves for the adjustment of the platinum mixing sphere. Complete kinetic traces within a certain time range are recorded by a multichannel detector that contains a charge-coupled device (CCD). See also Teilum et al. (2002); Zhu et al. (2003).

continuous-flow observation cell. Using a flow cell (Fig. 5.3) instead of a free jet in air improves the optical stability, and the earliest observable time point in this simple design was still 100 ± 50 μs.

The continuous-flow capillary mixer with a 50-μs dead time (Fig. 5.4) utilized in Roder's group (Shastry et al., 1988; Sauder et al., 1996; Park et al., 1997; Shastry and Roder, 1998) combines the advantages of rapid mixing via a sphere and the good stability of a continuous-flow mixing cell. A multichannel charge-coupled device (CCD) detector enables the real-time recording of complete kinetic fluorescence traces without moving the detector. Different channels of the CCD correspond to different time points. Simultaneous detection of the complete kinetic traces within a certain time range of typically microseconds to milliseconds significantly speeds up the measurement and improves optical stability.

Ultrafast continuous mixing has been combined with resonance Raman spectroscopy (Takahashi et al., 1997). Resonance Raman spectra of the heme group of cytochrome c suggest that in the folding reaction the protein is trapped in a misfolded conformation with two histidines ligated to the heme iron (Takahashi et al., 1997). The fraction of misfolded molecules has been significantly reduced by decreasing the pH from 5.9 to 4.5 which causes the protonation of the misligated histidines (Yeh et al., 1997). At pH 4.8 a rate constant for the main folding pathway of $1400 \ s^{-1}$ at 40°C has been observed. Further progress in the resolution of the folding kinetics of cytochrome c, using ultrafast mixing, has been made in William Eaton's and James Hofrichter's groups. Cytochrome c refolding has been studied in the previously inaccessible time range from 80 μs to 3 ms (Chan et al., 1997). Adding imidazole to the protein solution prevents misligation of histidines and dramatically speeds up folding (Chan et al., 1997).

Fig. 5.5 shows an ultrafast continuous-flow double-jump mixing device, made by the author, which is designed to extend the time scale of H/D exchange experiments monitored by NMR (see Sects. 7.1 and 8.1). The mixing tubes and the delay tube are made from grooves carved into a plate of acrylic glass which is sealed with a plate made from stainless steel.

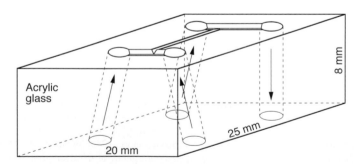

Fig. 5.5. Ultrafast continuous-flow double-jump mixing head with delay times of 200–500 μs. Two T-mixers are made from grooves carved into a plate of acrylic glass sealed on top with a plate made from stainless steel (not shown). The three inlet tubes and the outlet tube have diameters of about 0.5 mm. The device is attached to a common quenched-flow apparatus.

5.2
Temperature-jump

5.2.1
Electrical-discharge-induced T-jump

5.2.1.1
T-jump apparatus

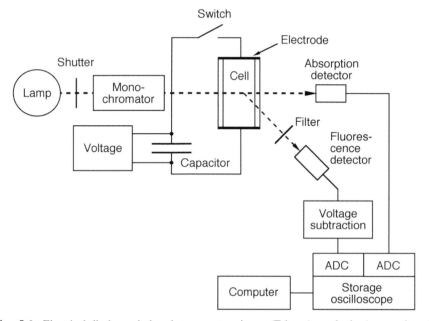

Fig. 5.6. Electrical-discharge-induced temperature-jump (T-jump) method. A capacitor is charged by a power supply up to a specific voltage and then rapidly discharged through the sample cell that contains the protein in a buffer with a certain electrical conductivity, for example, 50 mM phosphate buffer with 100 mM KCl. The electrical discharge causes Joule heating by 1–20°C with rise times of typically 500 ns – 10 μs, depending on the instrument settings, in particular on resistance of the protein solution and capacitance. When starting from the (partially) cold-unfolded state, increase of temperature causes refolding, otherwise fast unfolding reactions may be studied. The reaction kinetics is followed by absorption or fluorescence detection. The electrical signals are digitized by analog-to-digital converters (ADC) which are part of an Nicolet (Madison, WI) model Pro 90 storage oscilloscope and further processed on a computer. Large sample cells with 1 mL volume may be used which enables a high light throughput and thus an excellent signal-to-noise-ratio (see Fig. 5.8). Minimization of photolysis in the sample by the intense light of a 200-W mercury–xenon lamp is achieved with the help of an optical shutter that opens only during the measurement. Prior to digitalization of the fluorescence signal by an ADC, a constant voltage is subtracted (see Fig. 5.9). Therefore, a 12-bit ADC is usually sufficient to resolve changes of only 0.01% in the nanosecond time scale. When using a well-stabilized power supply for the lamp, usually no reference channel is needed. Otherwise the signal of a reference detector that is located between monochromator and sample cell (not shown) may be used. The simplicity of the device that does not contain mechanically moving parts makes the handling very easy and causes an exquisite reproducibility.

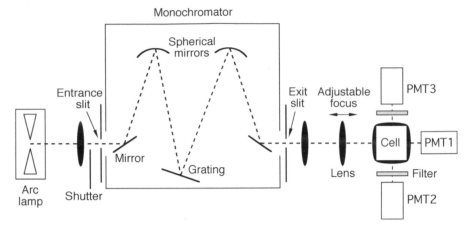

Fig. 5.7. Optics of an electrical-discharge-induced T-jump apparatus (Fig. 5.6). Photomultiplier tube 1 (PMT1) is used for absorption measurements. For fluorescence detection, the electrical signals of photomultiplier tubes 2 and 3 (PMT2 and PMT3) are added. To decrease the photon shot noise, the optics is optimized for a large light throughput (e.g., relatively wide bandwidths of fluorescence excitation and emission are used for the experiments presented in Chap. 10) and a high aperture of fluorescence detection (DIA-LOG, Düsseldorf, Germany; Eigen and deMaeyer, 1963; French and Hammes, 1969; Nölting et al., 1995, 1997a).

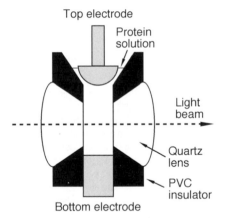

Fig. 5.8. Sample cell for an electrical discharge T-jump apparatus. In order to avoid pressure due to thermal expansion upon T-jump, the top of the cell is not sealed. Fluorescence detection is perpendicular to the excitation beam (DIA-LOG, Düsseldorf, Germany; Eigen and deMaeyer, 1963; French and Hammes, 1969; Nölting et al., 1995, 1997a).

Protein solutions that contain electrolytes may have sufficient electrical conductivity to enable them to be heated by a rapid electrical discharge through the sample cell (Figs. 5.6–5.10). With the simple design (Eigen and deMaeyer, 1963; French and Hammes, 1969) Joule heating with rise times of ≈1 µs or faster can easily be achieved when using a buffer with 100 mM KCl. The large size of the sample cell of about 1 mL volume enables a large light flow and thereby leads to a low photon shot noise. Noise levels of <0.01% root mean square (rms) of the fluorescence signal have been achieved at a 5-µs response time of the electronics (Nölting et al., 1995). The amplitude of temperature-jump (T-jump), ΔT, is given by

$$\Delta T \approx \frac{CU^2}{\rho V c_p} \ , \tag{5.1}$$

where ρ, V, c_p, C, and U represent density, volume and specific heat capacity of the sample, electrical capacitance, and applied voltage, respectively. The exponential rise time, τ, of temperature is approximately

$$\tau = RC \ , \tag{5.2}$$

where the resistance, R, and capacitance, C, typically are around 50–200 Ω and 10–50 nF, respectively (Fig. 5.10).

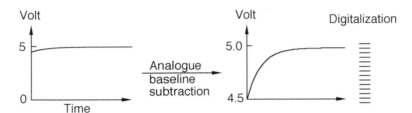

Fig. 5.9. Improvement of the detection of small changes of fluorescence in T-jump experiments: Prior to the T-jump, the signal is adjusted to zero by subtracting a constant voltage. Thereby faster analog-to-digital converters (ADC) with lower digital resolution may be used.

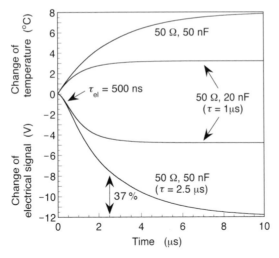

Fig. 5.10. Electrical-discharge-induced T-jumps with different instrument settings. The measured electrical signal was derived from the "instant" change of protein fluorescence upon temperature change. When adjusting the resistance of the sample to 50 Ω and using a 50-nF capacitor, the rise time, τ, is 2.5 μs. With a 20-nF capacitor, the rise time can be shortened to 1 μs. However, at the same voltage, the amplitude of the T-jump is 2.5 times smaller. The electrical signal has a slightly different shape due to the response time of the electronics, in this example 500 ns.

In general, a T-jump on a protein perturbs the equilibrium constant of the folding reaction, K (Eigen and deMaeyer, 1963; French and Hammes, 1969):

$$\left(\frac{\partial \ln K}{\partial T}\right)_P = \frac{\Delta H}{RT^2} ,$$

(5.3)

where ΔH is the enthalpy change of the reaction, $R = 8.314$ J mol^{-1} K^{-1} is the molar gas constant and T the absolute temperature. For small-amplitude jumps, ΔT, follows:

$$\left(\frac{\Delta K}{K}\right)_P = \Delta T \frac{\Delta H}{RT^2} .$$

(5.4)

5.2.1.2
Observation of early folding events: refolding from the cold-unfolded state

Even though temperature-jumping (T-jumping) has been used for a long time for the observation of spin-relaxations of heme-proteins (Sligar, 1976; Fisher and Sligar, 1987), of fast protein unfolding (Tsong et al., 1971; Lin and Cheung, 1992), and of fast conformational relaxations in proteins (Eigen et al., 1960; Cathou and Hammes, 1964, 1965; Wang et al., 1975; Jentoft et al., 1977; Feltch and Stuehr, 1979; Tsong, 1982; Steinhoff et al., 1989; Walz, 1992; Narasimhulu, 1993), the application to protein refolding was complicated by the fact that temperature increase usually favors loss of structure, i.e., unfolding (Phillips et al., 1995).

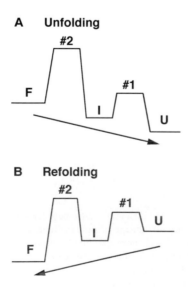

Fig. 5.11. Observation of early (re)folding intermediates. Energy landscape of a protein that contains a high (#2) and a low (#1) transition state, and an intermediate (I) on the pathway between unfolded (U) and folded (F) state. **A:** In unfolding experiments the early (re)folding intermediate cannot easily be observed since its population is always low. This is because the intermediate state slowly becomes populated and rapidly depopulated. **B:** Refolding experiments allow us to observe intermediates that are located on the reaction coordinate between the main transition state, #2, and the unfolded state: The amplitudes of both kinetic events are high.

However, early folding events can hardly be resolved in unfolding experiments (i.e., under conditions that favor unfolding). This is illustrated in Fig. 5.11, which displays the energy landscape of a protein that has an early folding intermediate, I, which is located on the reaction coordinate in-between the unfolded, U, and the main transition state, #2. One can see that in unfolding experiments (Fig. 5.11, A) the molecules slowly pass through the high transition state barrier, #2, and thus the intermediate state slowly becomes populated. Then the molecules rapidly pass through the low-energy transition state, #1, thereby rapidly de-populating the intermediate state. Thus, the occupancy of the intermediate state is always low, and the fast transition from I to U may not easily be monitored since it has a very small amplitude. In contrast, in refolding experiments (Fig. 5.11, B), after passing through the fast transition, the molecules become trapped by the high transition state barrier which can only slowly be overcome. Thus, both phases have a high amplitude and may be measured (see also Sect. 4.3.1.1).

Fortunately, globular proteins usually display the property of cold-unfolding (cold-denaturation), i.e., a decrease of stability with decreasing temperature (Fig. 5.12; Privalov, 1990; Griko and Privalov, 1992; Damaschun et al., 1993; Gast et al., 1993, 1995; Nishi et al., 1994). The temperature-dependence of the Gibbs free energy change upon folding, ΔG_{F-U}, is

$$\Delta G_{F-U} = \Delta H_{F-U,g} \left(1 - T/T_g\right) + \Delta C_p(T - T_g - T \times \ln(T/T_g)) , \qquad (5.5)$$

where T, T_g, $\Delta H_{F-U,g}$, and ΔC_p are the absolute temperature, the absolute temperature at the midpoint of the heat-unfolding transition, the change of enthalpy at T_g, and the change of molar heat capacity, respectively (Pfeil, 1988).

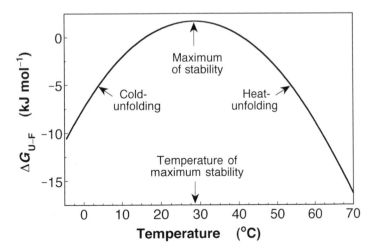

Fig. 5.12. Thermodynamic stability, $\Delta G_{U-F} = -\Delta G_{F-U}$, of P27A/C40A/C82A barstar in 2 M urea as function of temperature, calculated from data in (Nölting et al., 1995). Qualitatively most other globular proteins show a similar behavior, where the stability increases upon cooling at high temperatures, but upon heating at low temperatures.

At first glance, this surprising thermodynamic phenomenon might be counter-intuitive since one might expect a higher degree of disorder with increasing temperature but not the opposite. The reason for the existence of cold-induced unfolding of protein is seen in the fact that the protein is dissolved in water, and the overall entropy of the system represents a delicate balance of the contributions from water and protein (Makhatadze and Privalov, 1993; Weber, 1996): Upon lowering the temperature, the structure-forming propensity of water increases; the hydrophobic stabilization of the protein core decreases, and the preference of polar groups in the core for hydration increases (see Sect. 3.4). Thus, it has been realized that rapid refolding may be triggered in T-jump experiments when starting from the cold-unfolded state (Nölting et al., 1995; Nölting, 1996).

Probably all globular proteins cold-unfold at a sufficiently low temperature, but for many proteins in the absence of denaturants, a significant degree of cold-unfolding can only be achieved at temperatures below the freezing point of water. In these cases, the experiments are performed at different denaturant concentrations, and the kinetic rate constants are extrapolated to zero concentration of denaturant. Thus, the T-jump method for the rapid initiation of refolding is generally applicable to a wide range of proteins.

A protein of enormous interest has been barstar, the 10 kDa inhibitor of the ribonuclease barnase, for which the occurrence of an early intermediate has been predicted from burst-phase analysis (Schreiber and Fersht, 1993b; Shastry and Udgaonkar, 1995). NMR and CD spectroscopy have shown that the cold-denatured state of barstar is a highly unfolded state with only a small amount of residual structure, especially in $helix_1$ and $helix_2$ (Wong et al., 1996; Nölting et al., 1997b). T-jumping of cold-unfolded barstar has enabled the first kinetic and structural resolution of the folding pathway of a protein at the level of individual amino acid residues from microseconds to seconds (Chap. 10; Nölting et al., 1995, 1997a). Early on the folding pathway, barstar passes through an intermediate with a rate constant of formation of 2300 s^{-1} at 10°C and rate constant of decay of 800 s^{-1}. The activation enthalpy of the fast event is positive and its fluorescence change accounts for roughly 40% of the total for the transition from the unfolded to the folded state in the absence of denaturants. Φ-value analysis (see Sect. 8.3) shows that a significant degree of secondary and tertiary structure, mainly located in $strand_1$, $helix_1$, and $helix_4$, is formed on a time scale of ≈ 1 ms. At this stage, large parts of the molecule are still in a molten-globule-like state. The high structural resolution of the folding pathway of barstar from microseconds to seconds (Nölting et al., 1995, 1997a; Nölting, 1998a) reveals that the folding mechanism is consistent with a nucleation–condensation model (Abkevich et al., 1994b; Itzhaki et al., 1995a; Fersht, 1995c, 1997; Freund et al., 1996; Ptitsyn, 1998; Nölting, 1999). Similar to the growth of a crystal, a part of the molecule, the so-called nucleus, forms early. The diffuse nucleus becomes increasingly stabilized as further structure condenses around it in the course of the folding reaction. Later folding events are characterized by a hierarchical assembly of structure (Chaps. 10 and 11; Nölting et al., 1995, 1997a; Nölting, 1998a, 2003).

5.2.1.3
Observation of unfolding intermediates

By using double-jumps, the T-jump method may be extended to the study of intermediates which are located on the reaction coordinate between the folded state and the main transition state (Nölting, 1996), so-called unfolding intermediates (Kiefhaber and Baldwin, 1995; Kiefhaber et al., 1995; Wallenhorst et al., 1997). Lactoglobulin which contains several prolines, unfolds on a time scale of hours in 4 M urea at temperatures around −4°C. When T-jumping after a short incubation time under conditions of partial cold-unfolding, lactoglobulin displays a rapid kinetic event with increase in fluorescence intensity (Nölting, 1996). The amplitude of this kinetic event disappears with the same rate constant as the protein approaches the fully unfolded state, suggesting the occurrence of an unfolding intermediate (Figs. 5.13, 5.14; Nölting, 1996).

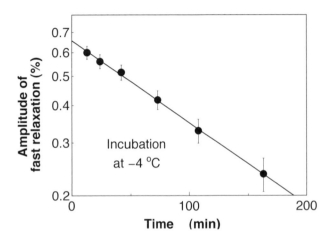

Fig. 5.13. Refolding of β-lactoglobulin from the partially cold-unfolded state: The amplitude of the fast reaction disappears after prolonged incubation under conditions of cold-unfolding at −4°C in 4.5 M urea (Nölting, 1996).

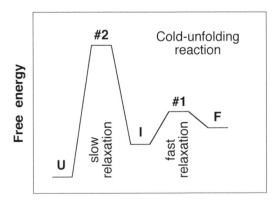

Fig. 5.14. Free energy diagram of β-lactoglobulin at low temperatures. U, I, F, unfolded, intermediate, and folded states; #2 slow transition state; #1 fast transition state (Nölting, 1996).

5.2.2
LASER-induced T-jump

When the electrical discharge was replaced by a LASER pulse, temperature rises in the nanosecond time scale were obtained (Fig. 5.15; Thompson, 1997).

At wavelengths around 1.5–2 micrometers the absorption of H_2O and D_2O is not too large to ensure a relatively uniform heating in a sample cell of about 50 micrometers path length. Nanosecond pulses of several mJ at this wavelength may be generated using optical parametric oscillators (OPO; e.g., from B.M.Industries, Lisses, France; Elliot Scientific, Harpenden, U.K.; Continuum, Santa Clara, CA; see Fig. 5.16), or even stronger pulses with up to about 200 mJ using Raman shifters (e.g., Edinburgh Instruments, Edinburgh, U.K.; Lambda Photometrics, Harpenden, U.K.; Light Age, Somerset, NJ; see Fig. 5.17; Ballew et al., 1996b). Temperatures around 4°C where the coefficient of thermal expansion of water approaches zero, small sample cell volumes, and small amplitudes of the T-jumps are used to avoid significant pressure effects upon rapid heating.

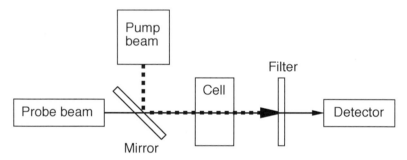

Fig. 5.15. LASER T-jump and optical triggers. One LASER beam serves as a pump beam to excite the desired chemical or physical changes which are measured by using a probe beam.

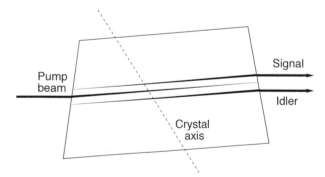

Fig. 5.16. Optical parametric oscillator. The nonlinear polarizability of the crystal medium is used to generate two low frequency photons (the signal and idler waves) from a single, high frequency photon (pump wave): 1/pump wavelength = 1/signal wavelength + 1/idler wavelength.

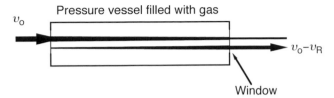

Fig. 5.17. Raman shifter. The pump line with the wavenumber v_0 (1/wavelength) is converted (scattered) into the Stokes line $v_0 - v_R$. The magnitude of v_R depends on the gas. For example, v_R is about 4155 cm^{-1}, 2991 cm^{-1}, 2331 cm^{-1}, and 2915 cm^{-1}, for H_2, D_2, N_2, and CH_4, respectively. Optimal conversion efficiencies at pressures of several 10 bar (several MPa) are typically 10–50% at the first Stokes wavelength. Anti-Stokes lines and higher-order Stokes lines have lower intensities. For better visibility, the optical axes of the two beams are shifted against each other.

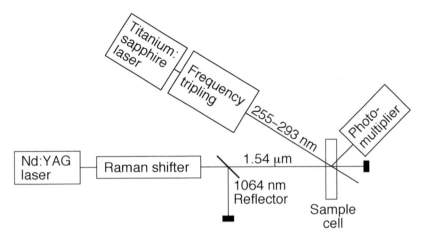

Fig. 5.18. Nanosecond T-jump machine with real-time fluorescence detection. 1.54-μm light pulses with energies of 200 mJ are generated by passing the 1.064-μm output of a Q-switched Nd:YAG LASER through a Raman shifter (see Fig. 5.17). Temperature-jumps in the sample cell are achieved by utilizing water absorption at 1.54 μm. The frequency-tripled output of a titanium–sapphire LASER serves for the excitation of protein fluorescence which is used to monitor the reaction kinetics initiated by the temperature change. When T-jumping the protein from the cold-unfolded to the folded state, similar to that described in Sect. 5.2.1.2, early folding events in the nanosecond time scale may be detected (Service, 1996; Ballew et al., 1996a, b, c).

The folding reaction proceeds under essentially isothermal conditions, since the temperature of a typical protein molecule equilibrates with the solvent within nanoseconds or faster depending on the molecular weight (see Sect. 5.2.3; Nölting, 1995; Ballew et al., 1996a; Nölting, 1998b).

Several groups have applied the method (Nölting et al., 1995) of T-jumping of protein from the (partially) cold-unfolded state. Using a T-jump machine with real-time fluorescence detection in the nanosecond time scale (Figs. 5.18, 5.19), Martin Gruebele's group demonstrated the partial formation of a subunit of myoglobin, comprising the A, G, and H helix of the molecule, in an initial folding event of only about 7 μs duration (Service, 1996; Ballew et al., 1996a, b, c; Sabelko et al., 1998). Since methionine 131 is a significant quencher of fluorescence in the folded state, folding at this position is connected with a decrease of the fluorescence signal. More work on myoglobin folding was also performed by other groups (Gilmanshin et al., 1997a, b, 1998; Pappu and Weaver, 1998).

The detection method used in the LASER T-jump set-up has been extended to far-infrared absorption at wavenumbers around 1500–1700 cm^{-1} (5.88–6.67 μm wavelength), which is a very sensitive probe of the protein secondary structure content. Helix–coil transitions of a 21-amino acid residue peptide with a folding rate constant of 6×10^7 s^{-1} at 28°C have been observed directly in the time domain (Williams et al., 1996; Thompson et al., 1997). The folding transition of a hairpin consisting of 16 amino acid residues is surprisingly slow, ≈6 μs at room temperature (Muñoz et al., 1997, 1998).

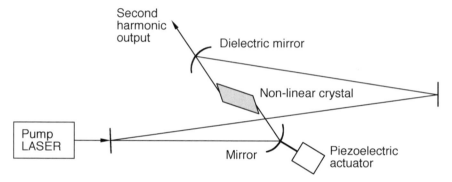

Fig. 5.19. Second-harmonic generation in a non-linear crystal which is contained in a ring-cavity. Short wavelengths useful for the excitation of fluorescence or for the measurement of protein circular dichroism are generated by doubling the frequency of a pump LASER.

5.2.3
Maximum time resolution in T-jump experiments

When heating the buffer via electrical discharge or LASER, the maximum speed of initiation of rapid folding is usually limited by the speed of the propagation of heat in the sample. An approximate analytical solution for the heat diffusion in and around a spherical molecule that is contained in a constant energy flow has been given, originally in the context of sound velocity measurements (Nölting, 1995), and later extended to T-jumping (Nölting, 1998b). Fig. 5.20 shows different temperature profiles in and around the protein molecule as a function of the parameter $\xi = (2\Lambda t \rho^{-1} c_p^{-1} r_o^{-2})^{0.5}$, where Λ, ρ, c_p, r_o and t are heat conductivity, density, specific heat capacity, radius of the molecule and time, respectively

(Nölting, 1995, 1998b). For $\xi \gg 1$, corresponding to a small radius of the molecule and a large time, there is a good temperature equilibration between protein molecule and bulk solvent.

This evaluation is in good agreement with LASER T-jump experiments in which the buffer is heated via excitation of a dye (Phillips et al., 1995). From experiment and theory it is estimated that the initiation of protein folding via T-jumping may be extended down to about 20 ps for a 10-kDa protein (Nölting, 1998b).

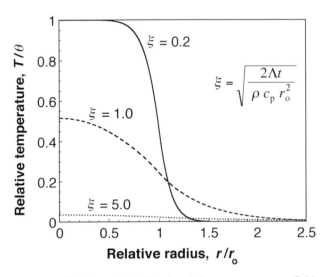

Fig. 5.20. Temperature profile for a spherical molecule in a constant energy field as a function of the parameter ξ, where Λ, ρ, c_p, r_o, t, and T are heat conductivity, density, specific heat capacity, radius of the molecule, time, and temperature change, respectively. r is the radius under consideration, and θ is the temperature change which would be obtained in the absence of heat conduction (Nölting, 1998b).

5.3
Optical triggers

In devices for optical triggers, a LASER pump beam usually causes a desired chemical change and a probe beam tests the induced changes (Fig. 5.15). The wavelength of the pump beam is selected to excite specific chromophores of the protein or of attachments to the protein, rather than water in T-jumping.

5.3.1
LASER flash photolysis

Flash photolysis was already being used in 1993 in groundbreaking experiments on cytochrome c and the structural resolution of refolding in the nanosecond time scale was obtained (Jones et al., 1993). Since the CO-bound cytochrome c is less

stable than cytochrome c, refolding of the protein may be initiated by flash
photolysis of CO. LASER flash photolysis has been used for a long time in
protein unfolding experiments and for the observation of small conformational
relaxations in proteins (Dyer et al., 1989; Hofrichter et al., 1991; Xie and Simon,
1991; Causgrove and Dyer, 1993; Hu et al., 1996). But the method developed in
Eaton's group enabled the first ultra-rapid refolding. Photodissociation takes less
than 100 femtoseconds (Boxer and Anfinrud, 1994; Eaton et al., 1996b; Figs.
5.21, 5.22). A thorough and precise analysis of the early folding events shows a
complicated refolding behavior where rebinding of native and non-native heme
ligands occurs. Rebinding of carbon monoxide limits the applicability of the
method to the fast time scale.

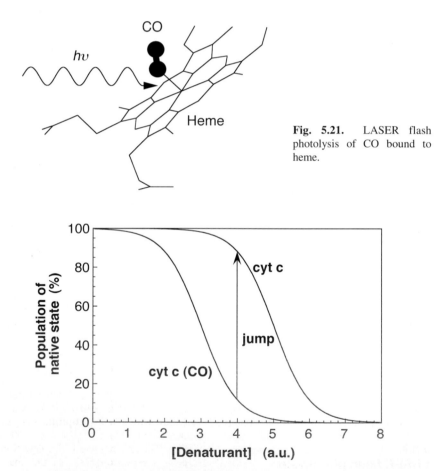

Fig. 5.21.   LASER flash
photolysis of CO bound to
heme.

Fig. 5.22. LASER flash-photolysis-induced refolding of cytochrome c. CO–cytochrome c is
less stable than cytochrome c. Flash photolysis of CO–cytochrome c in moderate concentrations of denaturant initiates rapid refolding (Jones et al., 1993).

Infrared detection of conformational changes following protein–CO dissociation with a time resolution of a few nanoseconds or less became possible due to the availability of infrared LASERs (e.g., diode LASERs manufactured by Laser Components, Olching, Germany) and fast mercury cadmium telluride (MCT) detectors (e.g., made by Kolmar Technologies, Conyers, GA; EG&G Judson, Montgomeryville, PA; Fermionics, Simi Valley, CA) with high sensitivity around 6 µm wavelength (Fig. 5.23; Dyer et al., 1994; Yuzawa et al., 1994).

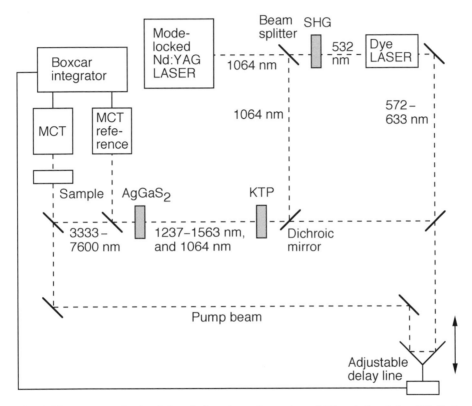

Fig. 5.23. An example for infrared detection of protein refolding induced by CO flash photolysis. The output of a Nd:YAG LASER (1064 nm) is frequency-doubled (532 nm) by using second-harmonic generation (SHG) and converted into a wavelength of 572–633 nm with a dye LASER. In a first mixer made from a potassium titanyl phosphate crystal (KTP), the difference frequency between the 1064 nm and the output of the dye LASER is generated. The resulting difference frequency is mixed with 1064 nm in a second mixer made from silver thiogallate (AgGaS$_2$). This generates a pulse which is tunable from about 3333 to 7600 nm ($\approx$3000–1300 cm^{-1}). Mercury cadmium telluride (MCT) detectors are used for the detection of the infrared radiation. Strong pump and probe pulses of a few picoseconds duration may be obtained when using additional optical amplifiers. In this measurement principle, the time point of detection and the time resolution are determined by the position of the delay line and the duration of the LASER pulses, respectively. Thus, picosecond time resolution may be obtained even with slower MCT detectors (Dyer et al., 1989, 1994; Causgrove and Dyer, 1993).

Subpicosecond resonance Raman detection has been utilized for the detection of CO-flash-photolysis-induced conformational changes in hemoglobin and myoglobin (Fig. 5.24; Petrich et al., 1987; Varotsis and Babcock, 1993; Franzen et al., 1995). This measurement principle could be used also for protein refolding experiments in a similar way to that shown in Fig. 5.22.

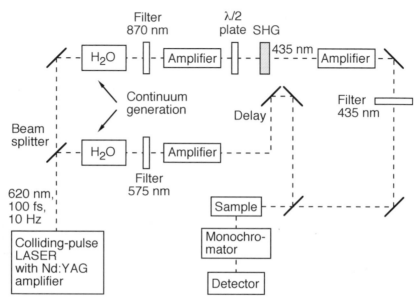

Fig. 5.24. Subpicosecond resonance Raman spectroscopy. A combination of a colliding-pulse LASER, continuum generation (H_2O), optical filters, second-harmonic generation (SHG), and pumped dye amplifiers is used to generate subpicosecond pulses of a pump beam at 575 nm and a probe beam at 435 nm (Petrich et al., 1987; Varotsis and Babcock, 1993; Franzen et al., 1995).

A further major breakthrough has been the extension of circular dichroism (CD) spectroscopy to the nanosecond time scale (Lewis et al., 1985; Lewis et al., 1992; Goldbeck and Kliger, 1993; Björling et al., 1996; Chen et al., 1998) by using a new type of CD spectrometer (see Sect. 8.2) in combination with LASER flash photolysis for the initiation of conformational changes in proteins. In this new spectrometer, the CD signal is optically amplified prior to the conversion into an electrical signal. Studies have been performed on kinetic processes effecting changes in the chiral structures of hemoglobin (Lewis et al., 1985; Björling et al., 1996), myoglobin (Lewis et al., 1985; Milder et al., 1988), and phytochrome A (Chen et al., 1997). Kliger et al. even succeeded in the measurement of the far-UV CD in the time scale from nanoseconds to seconds (Zhang et al., 1993; Chen et al., 1997; Chen et al., 1998).

A similar measurement principle with magnetic circular dichroism (MCD) detection (Milder et al., 1988) has been applied to conformational relaxations in cytochrome oxydase (Goldbeck et al., 1991; Woodruff et al., 1991), cytochrome ba_3 (Goldbeck et al., 1992), and cytochrome c_3 (O'Connor et al., 1993). MCD is sensitive to structural features of the protein molecule that directly affect the energy levels of the MCD chromophores (Goldbeck, 1988).

In a study by Chen et al. (1998) the refolding of cytochrome c, induced by flash photolysis of CO-bound cytochrome c in moderate concentrations of guanidine hydrochloride, was followed by CD in the far-UV (≥ 215 nm in this study), near-UV, and Soret region (about 400–460 nm). Conformational relaxations which have been observed in these experiments in the time scales of 500 ns and 2 µs are not connected with a significant formation of secondary structure (Chen et al., 1998). The method of time-resolved circular dichroism spectroscopy is presented in more detail in Sect. 8.2.

5.3.2
Electron-transfer-induced refolding

Rapid refolding of redox-active proteins for which there is a significant difference in stability between the oxidized and reduced forms has been induced by electron transfer (Chan et al., 1996; Mines et al., 1996; Pascher et al., 1996). The redox-active cofactor is covalently bound to the protein to prevent the effect of bimolecular reactions on the observed rate constant. Reduction of unfolded ferri cytochrome c within less than 1 µs initiates rapid refolding.

5.4
Acoustic relaxation

The frequency-dependence of sound absorption and sound velocity of protein is related to conformational transitions (Hammes and Roberts, 1969; Sarvazyan, 1991). Acoustic relaxation, i.e., sound absorption or sound velocity, can be measured using acoustic resonators (Fig. 5.25, 5.26; Eggers and Kustin, 1969; Gavish et al., 1983a, b; Sarvazyan, 1991; Chalikian et al., 1994, 1996) or by determining the time of flight of sound pulses (Sarvazyan, 1991). Electrical transducers which are used in acoustic resonators (Fig. 5.26) contain piezoelectric material that expands or contracts upon application of a voltage and can generate sound over the wide range of frequencies from ≤ 1 kHz to ≥ 1 GHz, corresponding to folding relaxations from ≥ 1 millisecond to ≤ 1 nanosecond.

Relaxations at about the frequency of the sound wave and at higher frequencies contribute to the frequency-dependent component of the sound velocity. The total observed signal is affected also by the properties of the bulk solvent, the properties of the hydration shell of the molecule, and the intrinsic properties of the molecule.

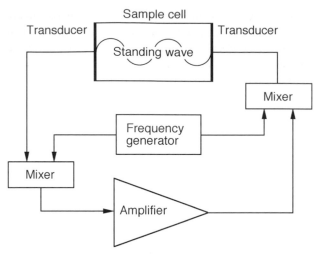

Fig. 5.25. Resonator sound velocity meter (Gavish et al., 1983a, b; Nölting et al., 1993; Nölting and Sligar, 1993). The resonator is composed of a sample cell that contains an emitting and a receiving transducer. A positive feedback-loop maintains a standing sound wave in the resonator. Since there is a number of resonance frequencies (higher harmonics), a frequency generator is used to select a specific oscillation. At a given number of nods of the standing wave, the resonance frequency is a linear function of the sound velocity. By precisely measuring the resonance frequency, the difference of sound velocity of the protein solution relative to buffer is determined with a precision of 0.2 cm s^{-1}, compared with the absolute sound velocity of water of 1500 m s^{-1} at 25°C ! The protein concentration is typically 1–10 g L^{-1}.

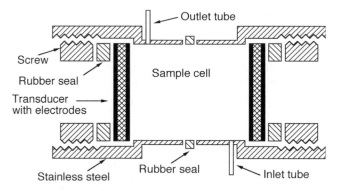

Fig. 5.26. Simple and robust design of a sample cell for a resonator sound velocity meter. For the precise coplanar adjustment of the transducers, the two sides of the cell are squeezed together with the help of three screws (not shown). The cell is temperature-controlled with a temperature drift of less than 0.005°C h^{-1}.

Significant acoustic relaxation is observed if there is a large volume change involved in the chemical or physical processes that are excited by the sound wave (see also Sect. 5.5). Computer dynamic simulations of the tiny volume fluctuations of the protein molecule are complicated because there are different definitions for the volume which is considered as the hydration shell, and so, the calculation of the volume of the protein molecule is ambiguous. Fortunately, the relaxational part of the acoustic signal can be identified by measuring at different frequencies with no need to separate hydration shell and intrinsic contributions.

Sound velocity, u, density, ρ, and adiabatic compressibility, β, of a solution are related to each other by the *Laplace* equation:

$$\beta^{-1} = \rho u^2 . \tag{5.6}$$

For the compressibility measurement of protein in solution by using Eq. 5.6, a high precision of density measurement is necessary, depending on the concentration of protein, which is typically $1–10$ g L^{-1} (for further information see Gavish et al., 1983a; Sarvazyan, 1991). Density meters for the measurement of small density differences with a precision of a few mg L^{-1} are commercially available (e.g., Mettler–Toledo, Greifensee, Switzerland).

Ultrasonic compressibility measurements have significantly expanded our understanding of the physical properties of molten globule states (MG). The MG of apo-cytochrome b_{562} at neutral pH displays only a small increase of relaxations at 2 MHz relative to the native conformation (Nölting and Sligar, 1993). Partially, this may be the case because in this MG three of its four helices have a relatively defined native-like secondary and tertiary structure. The sound velocity of the acidic MG of α-lactalbumin displays a large relaxational part at 2 MHz (Fig. 5.27; Nölting et al., 1993). Half of the sound absorption at 7 MHz is accounted for by relaxation (Kharakoz and Bychkova, 1997). This shows that a large amount of relaxations in the fluid-like MG state proceeds exceedingly fast, in the submicrosecond time scale.

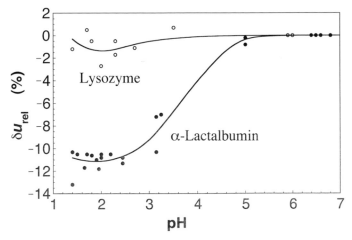

Fig. 5.27. Relative sound velocity changes $\delta u_{rel} = (\Delta u - \Delta u_o)/\Delta u_o$, where Δu_o is measured at pH 7, and Δu at the indicated pH. Closed circles, α-lactalbumin; open circles, lysozyme which has a large structural and sequential homology to α-lactalbumin (Nölting et al., 1993).

5.5
Pressure-jump

Pressure causes a perturbation of the physical or chemical equilibrium (Takahashi and Alberty, 1969). The relaxation of the system to a new equilibrium can be observed by an optical detection system (Fig. 5.28). Protein folding is generally accompanied by a change in volume. According to the principle of *LeChatelier*, an increase in pressure favors transitions to the state with the smaller volume (Weber, 1993; Jung et al., 1995, 1996; Topchieva et al., 1996), which is usually the unfolded state at sufficiently high pressure (Gross and Jaenicke, 1994; Foguel and Weber, 1995; Tamura and Gekko, 1995; Bismuto et al., 1996; Michels et al., 1996; Mozhaev et al., 1996; Tanaka and Kunugi, 1996).

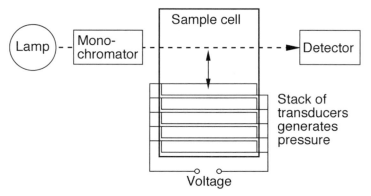

Fig. 5.28. Repetitive pressure-perturbation method (Pryse et al., 1992). A stack of transducers generates a pressure of 10–50 bar (1–5 MPa) which causes (partial) unfolding of the protein. After rapid release of the pressure by changing the voltage, the refolding kinetics of the protein is monitored by using absorption or fluorescence spectroscopy. Usually the stack of transducers is separated from the sample volume by a membrane (not shown).

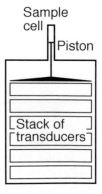

Fig. 5.29. Repetitive pressure-perturbation method (Pryse et al., 1992). Pressures of several 100 bar (several 10 MPa) may be generated using a smaller cross-section for the sample cell than for the transducers and a piston to propagate the force from the transducers to the sample cell.

The repetitive pressure-perturbation method (Figs. 5.28, 5.29) repeatedly uses pressure-jumps and accumulations of the kinetic traces (Pryse et al., 1992). Combined with signal averaging, the repetition of pressure-perturbations of about 10 bar (1 MPa) causes a dramatic improvement of signal-to-noise ratio and reproducibility, and a dead time of 1 ms has been obtained. Also in this method a further improvement down to a dead time of 100 µs should be possible, for example with smaller sample cells.

Single pressure-jumps of up to 700 bar (70 MPa) have been applied and relaxed within 60 µs using a mechanical valve (Fig. 5.30). Fluorescence detection revealed the slow refolding kinetics of barnase (Oliveberg and Fersht, 1996b).

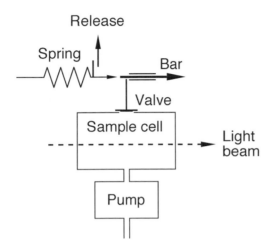

Fig. 5.30. A pressure-jump apparatus with a mechanical valve. After release of the spring, the bar is pushed which opens the valve. Fast pressure release enables jumps from 700 bar (70 MPa) to atmospheric pressure (Oliveberg and Fersht, 1996b).

5.6
Dielectric relaxation and electric-field-jump

Charge interactions are predicted to pull a protein apart if a sufficiently strong electric field is applied (DeMaeyer, 1969; Oliveberg and Fersht, 1996a). Early work on peptides using dielectric relaxation techniques (Fig. 5.31) has significantly contributed to our knowledge about the high speed of helix–coil transitions. It has been discovered that helix–coil transitions of small peptides may occur in the nanosecond time scale (Schwarz and Seelig, 1968).

The design of the field-jump set-up (Fig. 5.32) with a time resolution of 40 nanoseconds, which has been developed in the group of Manfred Eigen (Porschke and Obst, 1991; Porschke, 1996), resembles that of the electrical discharge T-jump apparatus illustrated in Figs. 5.6–5.8. The capacitor used in the T-jump apparatus is replaced by a coaxial cable. If the impedance of the cable is matched with that of the cell, the time of discharge is limited by the speed of the propagation of electromagnetic waves in the cable and by the length of the cable.

For example, the time of discharge of a 10-m cable is about 100 ns. Since the total released energy is comparatively small, the temperature increases are far smaller than in T-jumping. A low electrical conductivity of the protein solution is chosen, but the electrodes of the sample cell are not completely isolated from the protein solution because otherwise the electric field would rapidly be neutralized by counter ions from the sample. Similar to T-jumping but different from dielectric relaxation, the folding kinetics is monitored by fluorescence or absorption detection which enables the use of lower protein concentrations than those typically used in dielectric relaxation studies.

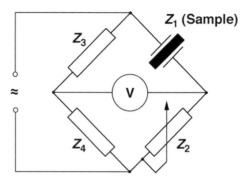

Fig. 5.31. Electrical wiring for the measurement of dielectric relaxation. An electrical bridge contains four resistors; one of them (Z_2) is adjustable; another (Z_1) represents the sample cell. In the measurement, Z_2 is adjusted to zero voltage at the meter indicated. Then Z_1 is given by: $Z_1 = Z_2 Z_3 / Z_4$. In the illustrated example of this measurement principle, Z_2 is a complex resistor that is constructed, e.g., by parallel connection of an adjustable capacitor with an adjustable Ohm resistor. Frequencies between 1 Hz and 1 GHz are relatively easy to generate and to apply.

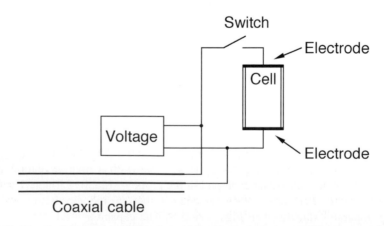

Fig. 5.32. Electric-field-jump machine. A coaxial cable is charged to a high voltage and rapidly discharged through the sample cell that contains the protein solution. The optical system (not shown) used for the detection of the fast kinetics is similar to that of the T-jump machine, see Figs. 5.6–5.8.

5.7
NMR line broadening

The principle of nuclear magnetic resonance (NMR) has been reviewed and discussed widely (see Sects. 7.1 and 8.1; Wüthrich, 1986; Williams and Fleming, 1995). Briefly, the effect of NMR originates from the absorption of electromagnetic radiation by the atomic nuclei which have a nuclear spin, such as ^{1}H, ^{13}C, and ^{15}N, when a magnetic field is applied. Nuclei with different

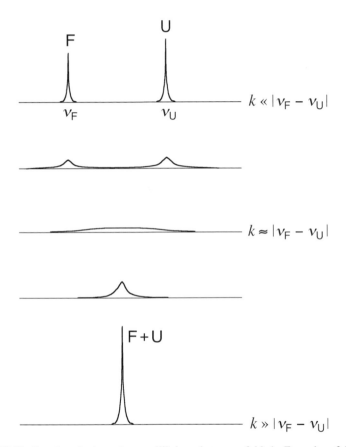

Fig. 5.33. NMR line broadening. An equilibrium between folded, F, and unfolded, U, conformations is considered, where k is the rate constant of exchange between U and F. v_F and v_U are the NMR frequencies for a particular atomic nucleus in the folded and unfolded conformation, respectively. *Top:* Folded and unfolded conformations are in slow exchange, $k \ll |v_F - v_U|$. Separate lines are observed for the folded and unfolded states. *Middle:* Intermediate frequency of exchange, i.e., $k \approx |v_F - v_U|$. A complicated line shape, which varies with the magnitude of k, is observed and from this k may be calculated. *Bottom:* Folded and unfolded conformation are in rapid exchange, $k \gg |v_F - v_U|$. The atomic nucleus "feels" only an averaged environment, and thus only one line is observed.

orientations relative to the externally applied magnetic field differ in their energy and in their occupancy. Transitions between these orientations may be excited by electromagnetic radiation.

Usually, an NMR line changes its position by Δv ($= |v_F - v_U|$) upon transition from the folded, F, to the unfolded state, U, because of a change of the chemical environment of the nuclei. Under conditions in which F and U are in equilibrium, three cases may be distinguished (Fig. 5.33; Williams and Fleming, 1995):

1. The rate constant, k, of the transition between F and U is low relative to Δv: Two lines are observed. The two amplitudes correspond to the populations of F and U, respectively (Fig. 5.33, top).
2. The rate constant of the transition between F and U is high relative to Δv: Only a single line is observed, since the nucleus "feels" only an averaged environment, in-between that of F and U (Fig. 5.33, bottom).
3. The rate constant of the transition between F and U is comparable to Δv: A complicated lineshape, which strongly varies with the magnitude of k, is observed. From this shape, the rate constant of exchange can be calculated for the specific nucleus (Fig. 5.33, middle).

With high-field NMR spectrometers, rate constants of exchange between F and U in the submillisecond time scale may be estimated using line broadening measurements of suitable ^{1}H lines with changes of chemical shifts upon unfolding by a few ppm (10^{-6}). NMR has the highest structural resolution among the methods mentioned in this chapter, but it requires a significant population of the species involved, and thus, is difficult to apply to early intermediates of low occupancy in equilibrium.

The aromatic ^{1}H NMR spectra of a truncated form of the N-terminal domain of phage λ-repressor have been measured at various concentrations of urea from 1.3 to 3.1 M (Huang and Oas, 1995). The extrapolated folding rate constant for the absence of denaturants is (3600 ± 400) s^{-1} at 37°C !

5.8
Summary

Currently, the highest, i.e., atomic, resolution of protein folding events may be achieved by NMR-related methods (Table 5.1). Usually, sample concentrations of 0.1 to 5 mM are required to obtain a sufficient NMR signal, and thus systems which are prone to strong aggregation may not be investigated. Currently, the time resolution of NMR when combined with H/D exchange (Sect. 8.1) initiated by ultrafast mixing (Sect. 5.1) is roughly 200 µs.

Significantly higher time resolution at low protein concentration has become possible along with the development of a number of fast methods for the initiation of folding, including temperature-jumping, optical triggers, acoustic relaxation, pressure-perturbation, and dielectric relaxation. With the exception of acoustic and dielectric relaxation, these methods may be combined with the Φ-value analysis which enables structural resolution at the level of individual amino acid residues, and is the only existent method to characterize structurally the transition states of folding (see Sect. 8.3).

Table 5.1. Methods for the study of fast folding events. For important results on ultrafast-folding proteins see Sect. 12.4.

Method	Approximate time range	Structural resolution
Mixing	$10 \ \mu s - \infty$	Atomic[a]
Electrical discharge T-jump	$100 \ ns - 10 \ s$[b]	Individual residues[c]
LASER T-jump	$1 \ ns$[d] $- 100 \ ms$[b]	Individual residues[c]
LASER flash photolysis	$100 \ fs$[e] $- 1 \ ms$	Individual residues[c]
Electron-transfer-induced refolding	$1 \ \mu s - 1 \ ms$	Individual residues[c]
Acoustic relaxation	$1 \ ns - 1 \ ms$	Global properties
Pressure-jump	$60 \ \mu s - 1 \ s$	Individual residues[c]
Dielectric relaxation	$1 \ ns - 1 \ s$	Global properties
NMR line broadening	$100 \ \mu s - 100 \ ms$	Atomic

[a] In combination with NMR, on a time scale of roughly $200 \ \mu s - \infty$.

[b] The resolution of slow kinetics is limited by the re-cooling of the sample after application of the T-jump, which depends on the size and heat conductivity of the sample cell. For example, the re-cooling rate constant of the sample cell with 1 mL volume shown in Fig. 5.8 is $0.03 \ s^{-1}$.

[c] Individual amino acid residues; in combination with protein engineering as described in Sect. 8.3 and Chap. 10.

[d] According to theoretical considerations, further extension to about 20 ps should be possible.

[e] The propagation of the conformational relaxation through the molecule may be slower.

6 Kinetic methods for slow reactions

6.1
Stopped-flow nuclear magnetic resonance (NMR)

One-dimensional NMR (see Sects. 5.7, 7.1, and 8.1) of small proteins with 5 to 20 kDa molecular weight may be performed on a time scale of a few 100 ms. This is fast enough to use stopped-flow techniques in combination with NMR to monitor folding reactions with rate constants of up to a few s^{-1} (Balbach et al., 1995; Hoeltzli and Frieden, 1995, 1996; Dyson and Wright, 1996). Fig. 6.1 displays a typical stopped-flow NMR design. The stop syringe used in common stopped-flow devices (Fig. 5.1) is replaced by stop bars. After a sample volume of a few 100 µL has passed through the mixer, the syringes are stopped and the acquisition of the NMR signal is started (Fig. 6.2).

One of the main applications of stopped-flow NMR is the structural resolution of folding intermediates. For higher sensitivity of stopped-flow NMR detection, the 5 tryptophan residues of *Escherichia coli* dihydrofolate reductase have been replaced by 6-^{19}F-tryptophan (Hoeltzli and Frieden, 1995, 1996, 1998). Using site-directed mutagenesis (see Sect. 8.3.1) the resonances have been assigned to individual tryptophan residues.

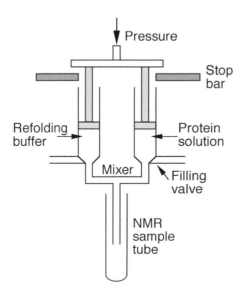

Fig. 6.1. Set-up for stopped-flow nuclear magnetic resonance. The stop syringe in the common stopped-flow apparatus (Fig. 5.1) is replaced by a stop-bar.

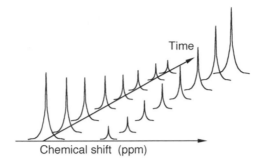

Time

Chemical shift (ppm)

Fig. 6.2. Change of NMR lines in stopped-flow-induced folding- or unfolding experiments (see text). Rate constants of the processes involved are determined by fitting the appropriate equations (see Chap. 4) to the intensities of the NMR lines as functions of time.

The folding pathway of dihydrofolate reductase has been found to involve the cooperative formation of one or more intermediates (Hoeltzli and Frieden, 1995, 1996, 1998). This direct NMR evidence for the cooperativity of folding represents a significant support of optical-spectroscopical and equilibrium-thermodynamical methods for the study of proteins in which the cooperativity of folding is often an important assumption for the interpretation of the data.

Complications in stopped-flow NMR experiments occasionally are transient and permanent aggregation (see Sect. 9.2) because the protein concentration needed is typically 100 µM to 5 mM. Permanent aggregation may easily be detected by using light scattering or ultracentrifugation methods (Sect. 9.2). However, transient aggregation, i.e., aggregation that occurs only for a short period of time after initiation of the reaction kinetics, may easily remain undetected, and may affect the properties of the molecules involved, or may even be confused with the formation of folding intermediates. However, NMR is not very sensitive to small populations of aggregates. Usually populations of less than 5% of aggregated or otherwise modified species are not detected.

6.2
Fluorescence- and isotope-labeling

Measurements of very slow folding kinetics or enzyme–inhibitor dissociation reactions may be problematical: Direct spectroscopic detection of the involved species may be too insensitive or not stable enough over weeks of data acquisition. In particular, fluorescence detection over a long period of time is complicated by photolysis of aromatic amino acid residues. In these cases the competition between two chemically similar but physically slightly different labels might be used for the observation of the reaction kinetics.

6.2.1
Folding reactions

1. A reaction mixture of protein and label$_1$ that will produce a high yield of protein–label$_1$ complex is made. Label$_1$ is an active label, i.e., it is detectable by optical spectroscopy, NMR, or radioactive methods.

2. The protein–label$_1$ complex is incubated under conditions of competition of label$_1$ with a large excess of the non-active label$_2$ that is chemically similar to label$_1$ but may be distinguished by physical methods. The conditions are chosen in such a way that label$_1$ may rapidly dissociate in the case of a partial opening or unfolding of the protein molecule (Fig. 6.3).
3. The concentration of the protein–label$_1$ complex is measured for different time points, in real-time or indirectly. Under suitable conditions, the rate constant of decrease of label$_1$ in the complex reflects the rate constant of partial opening or unfolding of the protein.

An important variant of this method is the quenched-flow technique (Sect. 8.1) which is used for the NMR- and mass-spectroscopical characterization of folding intermediates: Protein is labeled with ^{2}H that is later chased off by ^{1}H from the unprotected parts of intermediates. Under conditions of fast exchange, the rate constant of protection against label exchange reflects the rate constant of structure formation. ^{2}H (D, deuterium) is not radioactive and differs only moderately from ^{1}H in many properties, and thus the exchange often causes only a small perturbation of the sample. Usually the H/D exchange measurement is not real-time, and thus, may also be performed on a fast time scale.

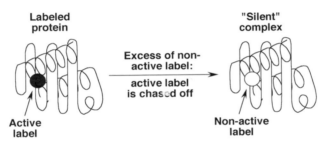

Fig. 6.3. Observation of very slow kinetics with the help of the method of competition between an active (i.e., detectable) and non-active label. Suitable labels may be fluorescent labels or isotopes. The active label is chased off with an excess of non-active label that is chemically similar. The rate constant for the decay of the concentration of the active complex is measured, e.g., by using optical methods, NMR, or scintillation counting. Under suitable conditions, the decay rate constant is equal to the rate constant of the global or local folding event.

6.2.2
Dissociation reactions

1. For the observation of the slow dissociation kinetics of a strong enzyme–inhibitor complex, enzyme–(inhibitor–label$_1$) complex and inhibitor–label$_2$ are made. Label$_1$ is detectable by optical spectroscopy, NMR, or radioactive methods. Label$_2$ is chemically similar but physically distinguishable from label$_1$, and may be a part of the inhibitor itself, for example, ^{1}H.
2. The enzyme–(inhibitor–label$_1$) complex is incubated under conditions of competition of the inhibitor–label$_1$ with a large excess of inhibitor–label$_2$.

3. Provided that the labels do not introduce significant perturbations of the reaction, the rate constant for the decrease in concentration of label$_1$ in the complex reflects the dissociation rate constant of the enzyme–inhibitor complex. This measurement under pseudo-first-order conditions is not significantly affected by the association rate because essentially every association event leads to the binding of the "silent" inhibitor–label$_2$ (Fig. 6.4).

With this method, the dissociation rate constant of the barnase–barstar complex has been determined using tritium (^{3}H) isotope labeling (Fig. 6.4; Schreiber and Fersht, 1993a, 1995). The barnase–barstar interaction is one of the strongest protein–protein interactions known. That is why the dissociation rate constant of this complex can hardly be measured by common mixing methods in which the rapid association interferes with the measurement of the dissociation rate constant.

In the presence of a large excess of native barstar, the concentration of ^{3}H-labeled barstar in complex with barnase decays single-exponentially, and the observed rate constant is equal to the dissociation rate constant of the complex. For the measurement of the disappearance of ^{3}H-labeled barstar from the complex, fractions of the sample were analyzed at different time points. For this purpose, the barnase–barstar complex was separated from the fraction of sample by using chromatography, and the concentration of ^{3}H in the complex was determined by using a scintillation counter (Schreiber and Fersht, 1993a, 1995).

Caution: ^{3}H is radioactive and must be used in strict accordance with all local and national regulations and laws.

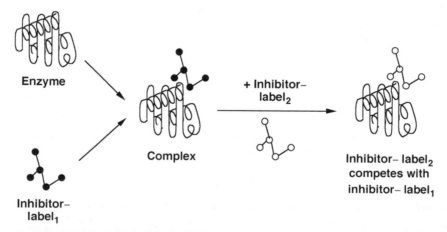

Fig. 6.4. Observation of very slow dissociation rate constants of an enzyme–inhibitor complex with the help of a competition method. Inhibitor–label$_1$ is chased off with an excess of inhibitor–label$_2$, and then removed, for example by using chromatography. For non-perturbing labels under suitable experimental conditions, the rate constant for the change of the concentration of enzyme–(inhibitor–label$_1$) complex reflects the dissociation rate constant of the enzyme–inhibitor complex (Schreiber and Fersht, 1993a, 1995). This method is analogously applicable to enzyme–substrate complexes.

7 Resolution of protein structures in solution

7.1
Nuclear magnetic resonance

Nuclear magnetic resonance (NMR) has emerged as an important tool for the study of protein structures, and it is the method with the highest structural resolution for proteins in solution that is currently available. A large number of excellent textbooks on protein NMR has been published (Atta-ur-Rahman, 1986; Wüthrich, 1986; Williams and Fleming, 1995). Extensive practical hints have been given by Croasmun and Carlson (1994). Here a brief introduction is presented to support the understanding of the kinetic applications of NMR in Sects. 5.7, 6.1, and 8.1.

NMR spectroscopy is based on the measurement of the absorption of electromagnetic radiation by the nuclei of atoms in the radio frequency range of several MHz up to several 100 MHz. The most important observable nuclei that occur in proteins or can be incorporated into proteins are ^1H, ^{13}C, ^{15}N, ^{19}F, ^{31}P. Their natural abundances are 99.985%, 1.10%, 0.367%, 100%, and 100%, respectively (Lide, 1993). Due to their low natural abundance, ^{13}C and ^{15}N often require isotope enrichment for a sensitive detection. For ^1H, ^{13}C, ^{15}N, ^{19}F, and ^{31}P the nucleic spin is 1/2, and thus, in a magnetic field the nucleus takes up two possible orientations that differ in the energy by ΔE (Williams and Fleming, 1995),

$$\Delta E = \gamma B \, h/(2\pi) \, , \tag{7.1}$$

where γ, B, and h are the magnetogyric ratio, the applied magnetic field strength, and Planck's constant, respectively. According to Boltzmann's distribution, the ratio of occupancy of the two orientations, N_β and N_α, is

$$N_\beta/N_\alpha = \exp(-\Delta E/(k_B T)) \, , \tag{7.2}$$

where k_B and T are Boltzmann's constant and absolute temperature, respectively. Electromagnetic radiation is absorbed when its frequency, v, matches the frequency at which the nuclear spins precess in the magnetic field (Wüthrich, 1986),

$$v = \gamma B/(2\pi) \, . \tag{7.3}$$

Eqs. 7.1 and 7.2 show that the occupancy difference, and thus the intensity of an NMR line, increases with increasing magnetic field strength, B. Furthermore, the spectral resolution also increases with increasing B.

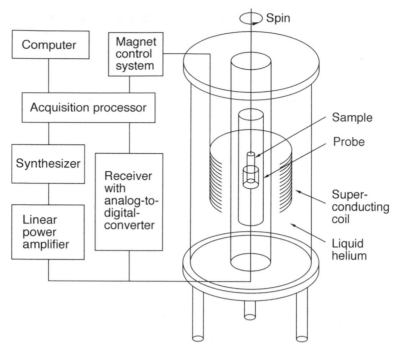

Fig. 7.1. Schematic representation of a high field Fourier transform NMR spectrometer. The sample is contained in a strong magnetic field (18.79 Tesla for an 800 MHz spectrometer) generated with a superconducting coil that is contained in liquid helium. Superconductivity is the ability of a material to carry electricity without any measurable resistance. The low temperature (typically 2–4 K) of the main coil enables a high magnetic field strength without breakdown of superconductivity. Spinning of the sample improves the homogeneity. Smaller coils contained in the probe serve for radiowave excitation and detection. Different pulse sequences are executed with the help of computer, acquisition processor, synthesizer and linear power amplifier. The free induction decay (FID) is amplified and digitized, and then further processed in the acquisition processor and computer (BRUKER Analytik, Karlsruhe, Germany).

Currently, NMR spectrometers with $B = 18.79$ Tesla are available (BRUKER Analytik, Karlsruhe, Germany). In high-field NMR spectrometers, the magnetic field is generated by using superconducting coils (Fig. 7.1). At a magnetic field strength of 18.79 Tesla, the protons (^{1}H) come into resonance at 800 MHz, and it is common to refer to an 800 MHz instrument.

The NMR approach to studies of biological macromolecules is based largely on four types of experiments: the chemical shifts, coupling constants, nuclear Overhauser effect (NOE), and amide proton exchange rates (Wüthrich, 1986; Williams and Fleming, 1995).

Chemical shifts. Since the magnetic field strength, B, is slightly affected by the chemical environment of the nucleus, i.e., electrons and other nuclei, the nuclei of the molecule come into resonance at slightly different resonance frequencies, υ. Lines are shifted to a different υ due to their different chemical environment. See, for example, Fig. 7.2 which displays the ^{1}H spectrum of ethanol: Resonance frequencies are measured relative to a reference frequency and are in units of ppm (part per million, 10^{-6}). Tetramethylsilane (TMS) is chosen as the internal standard, i.e., for the calibration of the reference frequency. One can see that the protons of the methyl group CH_3 (a) have the lowest chemical shift, i.e., lowest resonance frequency. These nuclei are significantly shielded by the electrons, and thus experience a lower magnetic field strength. The highest resonance frequency is found for the methylene protons (c) that are less shielded by electrons.

Coupling constant. Nuclei that are connected by covalent bonds will exert magnetic interactions on each other. The precession of covalently bound neighbored nuclei in the magnetic field affect each other, and fine structures, so-called multiplets, instead of single lines, are observed. This is called coupling. A nucleus that couples equally to n others will give rise to an $(n + 1)$-multiplet. For example, the signal of the methyl group, CH_3, of ethanol (Fig. 7.2) is a triplet caused by couplings of the methyl protons with the two methylene protons. Furthermore, the coupling constants vary with the conformation of the molecule (Williams and Fleming, 1995).

Nuclear Overhauser effect (NOE). NOE signals arise from spin–spin interactions of nuclei through space. These interactions influence the relaxation rates of nuclei and thereby affect the occupancies of different energy levels of the nuclei. The observed effect of these interactions is a change in intensity of an NMR line when another nucleus is irradiated at its resonance frequency. The NOE is only noticeable over a distance, d, of less than about 5 Å and its magnitude falls off as d^{-6}. Thus, NOE experiments may provide distance constraints between nuclei, which is very important information for simulating and refining the three-dimensional structure of a macromolecule.

Proton exchange rates. Proton exchange quenched-flow experiments are an important method for studying the rapidly changing conformations of macromolecules, such as protein folding intermediates (Sect. 8.1).

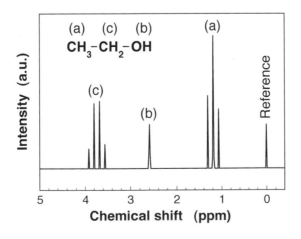

Fig. 7.2. ^{1}H NMR spectrum of ethanol.

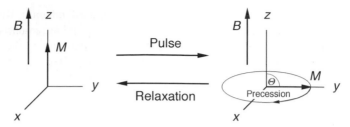

Fig. 7.3. Fourier-transform principle for NMR. The magnetization, M, of the sample is tipped through an angle, Θ, by a powerful radio frequency pulse. In this example $\Theta = 90°$ ($\pi/2$ - pulse). After the pulse has been applied, the magnetization starts to precess around the z-axis. The resulting radio frequency is detected and Fourier transformed (see Fig. 7.4). Relaxation of the magnetization causes an exponential decrease in the radio signal with time and return of the system to the original state.

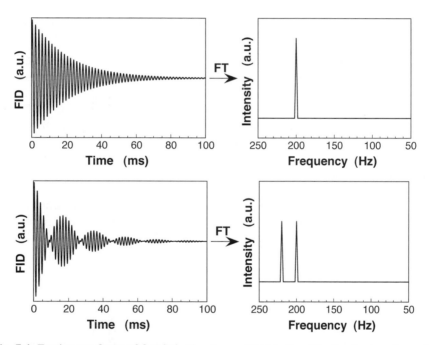

Fig. 7.4. Fourier transforms of free induction decays (FID's). *Top:* The Fourier transform of a single-sinusoidal function is a single NMR line. *Bottom:* A superposition of two sinusoidal functions in the time domain corresponds to two peaks in the frequency domain.

Precise data from mainly the first three of these four types of experiments serve for deducing the chemical environments of nuclei, for the assignment of the nuclei to specific amino acid residues, and for the simulation and the refinement of native protein structures. NMR spectroscopy on small stable proteins in solution can enable atomic resolution with errors of less than a few Å.

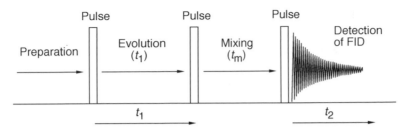

Fig. 7.5. Sequence of two-dimensional (2D) NMR experiments (see text).

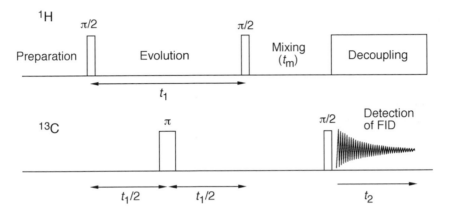

¹H magnetization:

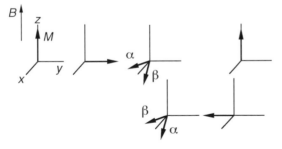

Fig. 7.6. Sequence for the heteronuclear two-dimensional nuclear Overhauser effect (HOESY). *Top:* Pulse sequences for protons and carbon nuclei. *Bottom:* Change of proton magnetization: The first 90° pulse tips the ¹H z-magnetization into the xy-plane. The ¹H magnetization vectors fan out during the first half of the evolution period ($t_1/2$). A 180° ¹³C pulse serves to interchange the spin labels and to refocus the ¹H magnetization vectors in the subsequent $t_1/2$ period. Another 90° ¹H pulse tips the magnetization back into the longitudinal direction. ¹H and ¹³C spins interact during the mixing period (Atta-ur-Rahman, 1986).

To obtain the NMR spectrum, the first spectrometers, so-called continuous wave spectrometers, used to measure the absorption while steadily changing the

radio frequency in the desired frequency range. Nowadays, most NMR spectrometers use a Fourier-transform principle: The radio frequency is applied as a
single pulse of typically a few µs duration. This pulse tips the magnetization of
the sample by a certain angle, Θ (Fig. 7.3). After application of the pulse, the
magnetization of the sample starts to precess around the z-axis. The difference
between the radio frequency caused by the precession and a reference frequency
is detected and Fourier transformed (Fig. 7.4). This mathematical operation transforms the signal from the time domain into the frequency domain. For example,
the Fourier transform of a sine function is a single line (Fig. 7.4 top). The superposition of two sine functions gives rise to a complicated interference in the time
domain (Fig. 7.4 bottom) but is easily recognized in the frequency domain.

A major breakthrough in the field has been the introduction of two-dimensional
(2D) NMR spectroscopy (Ernst and Anderson, 1966). Usually, the 2D NMR
measurement involves four phases that are cyclically passed through several times
(Figs. 7.5 and 7.6):

1. Thermal equilibration of the system is achieved during the preparation period.
   Similar to 1D NMR, a strong radio frequency pulse is applied which tips the
   magnetization by an angle Θ, for example $\pi/2$.
2. In contrast to 1D NMR, detection is not started immediately after this pulse, but
   the system is allowed to evolve, i.e., different nuclei are given time to interact.
   After a time of evolution of the system, t_1, a further radio frequency pulse is
   applied.
3. During the mixing period, t_m, the system evolves further.
4. Prior to the detection of the free induction decay (FID), a further pulse is
   applied in some experiments. The time of detection of the FID is t_2. Fourier
   transformation of the two time domains, t_1 and t_2, results in two frequency
   domains, ω_1 and ω_2, respectively.

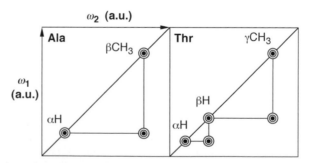

Fig. 7.7. Two-dimensional COSY NMR spectra of alanine (Ala) and threonine (Thr).

2D NMR enables us to determine the interactions between different nuclei and
thereby to estimate intramolecular distances and angles. To solve a protein structure, the first step is usually the assignment of the NMR peaks to the individual

amino acid residues. Different amino acid residues display various connectivities between the nuclei, and thus may be identified by using their different patterns of 2D NMR spectra, see, for example, Fig. 7.7. Further, different amino acid residues display different patterns of fine structure of the NMR peaks. After assignment of the NMR peaks to the individual amino acid residues, the information of the chemical shifts, the information from the connectivities, and most importantly, the information from through-space energy transfer experiments is used to establish distance constraints. Using the distance constraints and also energy potentials, the structure is simulated and refined. 2D and multidimensional NMR techniques provide not only information on the structures of folded proteins but also on folding intermediates, on unfolded states, and on the dynamics of the different conformations (e.g., Dyson and Wright, 1996; Eliezer et al., 1998). For further information see (Atta-ur-Rahman, 1986; Wüthrich, 1986; Croasmun and Carlson, 1994; Williams and Fleming, 1995).

7.2
Circular dichroism

Circular dichroism (CD) is exquisitely sensitive to changes of protein conformations (Johnson, 1990; Nakanishi et al., 1994; Duddeck, 1995). Far-UV CD, i.e., CD in the wavelength region of about 170–260 nm, is sensitive to the secondary structure, whereas the near-UV CD signal from about 260 nm to 330 nm is diagnostic of the chiral environment of aromatic residues and of the tertiary structure of proteins. CD signals above 330 nm may be sensitive to the conformation of cofactors, for example, heme. Although the resolution is not as high as with NMR (see Sect. 7.1), usually CD studies are much simpler and may be performed more cheaply and, most importantly, on a faster time scale (see Sect. 8.2).

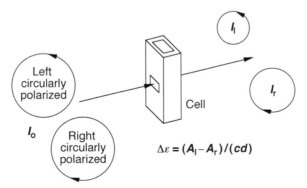

$$\Delta\varepsilon = (A_l - A_r)/(cd)$$

Fig. 7.8. Circular dichroism is the difference between the absorptions for left, A_l, and right, A_r, circularly polarized light. Alternately left and right circularly polarized light with the initial intensity I_0 is passed through the sample cell. In most CD spectrometers, the difference between A_l and A_r is calculated using solely the final intensities, I_l and I_r (see Fig. 7.9; Eq. 7.6; Velluz et al., 1965).

Protein CD is usually expressed in units of $\Delta\varepsilon$ or $\Delta\varepsilon_R$,

$$\Delta\varepsilon = \varepsilon_1 - \varepsilon_r \tag{7.4}$$
$$\Delta\varepsilon_R = \Delta\varepsilon / n_R ,$$

where ε_1 and ε_r are the molar extinction coefficients for left and right circular polarized light, respectively, and n_R is the number of amino acid residues of the protein (Fig. 7.8). $\Delta\varepsilon$ and $\Delta\varepsilon_R$ are called molar ellipiticity and mean residual ellipticity, respectively. The difference in absorption, $\Delta A = A_1 - A_r$, for left and right circularly polarized light is given by

$$\Delta A = \Delta\varepsilon \, c \, d , \tag{7.5}$$

where c and d are concentration of the sample and pathlength of the sample cell, respectively. Sometimes one still encounters the unit deg cm^2 dmol^{-1} ($[\theta]$) that is obtained by multiplying $\Delta\varepsilon$ with a factor of 3298.

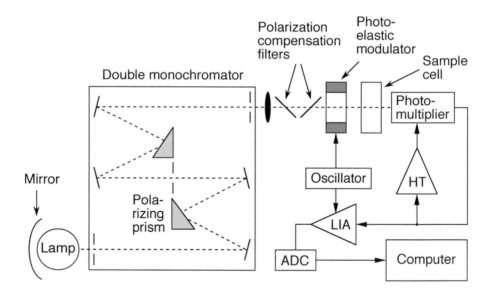

Fig. 7.9. Scanning CD spectroscopy. The light of a xenon lamp is passed through a polarizing double monochromator. Then the linearly polarized light is circularly polarized with the help of a modulator, passed through the sample cell, and converted into an electrical signal by a photomultiplier. The signal of the photomultiplier consists of an alternative current (AC) and a direct current (DC) component. A positive feedback loop ensures the constancy of the DC component by adjusting the high voltage supply of the photomultiplier (HT). The AC component with a frequency equal to the frequency of the modulator arises from the difference between the light absorptions with left and right circular polarization. Its amplitude is a measure of the circular dichroism. After phase-sensitive detection and rectification of the AC component by the lock-in amplifier (LIA), it is digitized by the analog-to-digital converter (ADC) and further processed in the computer. The CD spectrum is obtained by scanning the wavelength region of interest with simultaneous variation of the amplitude of modulation.

CD spectrometers are available for a wavelength range of about 165–1100 nm (e.g., from JASCO, Tokyo, Japan). In the widely used principle for scanning CD spectroscopy, a double monochromator ensures the selection of the desired wavelength and an excellent suppression of scattered light (Fig. 7.9). Circularly polarized light is generated by conversion of linearly polarized light with the help of a photoelastic or electro-optical modulator that acts as a retardation element with a variable phase delay (Fig. 7.10; Sect. 8.2). Photoelastic modulators made from SiO_2 are transparent to about 155 nm. For shorter wavelengths, CaF_2, MgF_2, or LiF_2 with a cut-off at about 130 nm, 125 nm, and 110 nm, respectively, are used (Nölting, 1991). For technical reasons, during the modulation cycles, most of the time the polarization is not exactly circular but elliptical. The CD is measured by passing alternately left and right elliptically polarized light through the sample and measuring the transmitted intensity. The difference in absorption, $\Delta A = A_l - A_r$, is calculated from the difference of the light intensities falling onto the detector,

$$\Delta A \approx \frac{2 \log(e)}{f} \frac{I_l - I_r}{I_l + I_r} , \qquad (7.6)$$

where I_l and I_r are the intensities of left and right elliptically polarized light, respectively, $2 \log(e) \approx 0.87$, and f is correction factor for the deviation of the polarization from exactly circular polarization. For the sinusoidal modulation in most CD spectrometers, f is about 0.79. The CD spectrum is obtained by scanning the wavelength region of interest. For this purpose, the prisms of the double monochromator are rotated with the help of computer-controlled step motors. For the measurement principle of CD and for the construction of CD spectrometers see also Velluz et al. (1965), Nölting (1991), and Sect. 8.2.

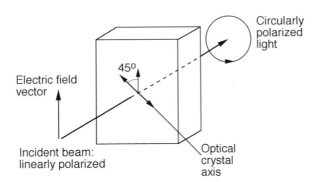

Fig. 7.10. Generation of circularly polarized light by passing linearly polarized light through a quarter-wavelength (λ/4) crystal phase plate. The angle of the optical axis of the phase plate relative to the plane of polarization of the incident beam is 45°. At this angle the crystal splits the incident beam into two components of exactly equal amplitude. These two linearly polarized components travel through the crystal plate with different velocities. The birefringence and thickness of the crystal are adjusted to cause a difference between the two components of exactly λ/4. The superposition of the two components of equal amplitude but 90° phase difference produces a circular polarization of the beam that exits the crystal.

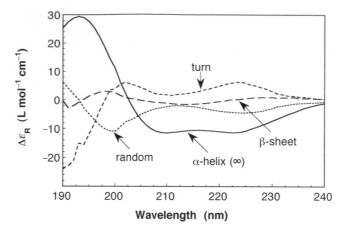

Fig. 7.11. Far-UV CD signal for different types of secondary structure: α-helix of infinite length, β-sheet, turns, and random coil, as indicated (Yang et al., 1986).

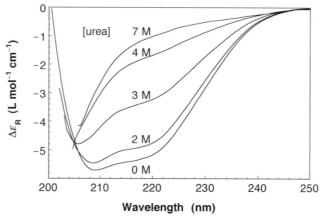

Fig. 7.12. Far-UV CD spectrum of a barstar mutant (C40A/C82A/P27A) at 25°C as function of the concentration of urea, as indicated (Nölting et al., 1997b). The protein unfolds upon increasing the urea concentration, [urea], which causes characteristic changes of the amplitude and shape of the CD spectrum around 220 nm.

Different types of secondary structure, α-helix, β-sheet, and turns, corresponding to different chiral environments of the amide chromophores, give rise to markedly distinct far-UV CD spectra (Fig. 7.11). The typical spectrum of barstar which is rich in α-helix is shown in Fig. 7.12. The precision of the determination of the secondary structure content by decomposition of CD spectra improves with increasing range of wavelengths used. However, below 180 nm the measurement of CD spectra is very complicated because of the absorption of protein, salts, and water, and the information content increases only slightly, considering that the methods of secondary structure determination by using decomposition of CD spectra have a considerable systematic error.

The sensitivity of the near-UV CD of proteins to changes of the tertiary structure content is illustrated in Fig. 7.13. Unfolding of barstar by urea, heat, or low temperature causes a weakening of the near-UV CD signal at 260–290 nm, and a change of sign, due to a loss of chiral environment of the aromatic sidechains. The signal at 295–310 nm that arises mainly from buried tryptophan sidechains vanishes almost completely upon heat-, cold-, or urea-induced unfolding (Nölting et al., 1997b).

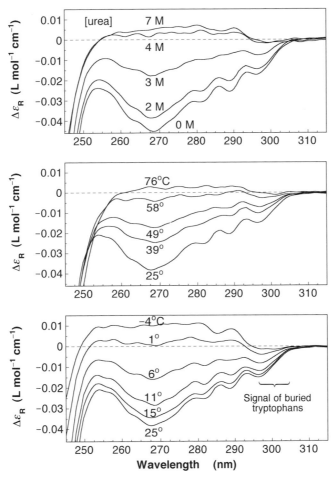

Fig. 7.13. Near-UV CD signal of a barstar mutant (C40A/C82A/P27A) under different conditions (Nölting et al., 1997b). *Top:* CD at 25°C as function of the concentration of urea, as indicated. *Middle and bottom:* CD in 2.2 M urea as function of temperature, as indicated.

8 High structural resolution of transient protein conformations

8.1
NMR detection of H/D exchange kinetics

Nuclear magnetic resonance (NMR) techniques not only monitor individual backbone amide protons, but also discriminate deuterons (D, ^{2}H) and protons (H, ^{1}H) (see also Sects. 5.7 and 7.1; Roder et al., 1988; Udgaonkar and Baldwin, 1988; Bycroft, 1990; Wüthrich, 1986; Williams and Fleming, 1995). Most of the amide protons of unfolded protein rapidly exchange with the protons of the solvent. Proton exchange is both acid- and base-catalyzed. For random chain polyalanine, the pD of slowest exchange is about 3 (Fig. 8.1; Englander and Mayne, 1992). For the fully folded protein the exchange rates of most of the amide protons are dramatically slowed down, in some cases by more than a factor of 10^9 (Udgaonkar and Baldwin, 1988). The protection against exchange of the amide protons originates from the burial of the amide hydrogens and from hydrogen bonding (Roder et al., 1988). Because folding intermediates usually already have protected parts in their structures, they may be characterized by using H/D exchange techniques in combination with NMR.

A typical quenched-flow H/D exchange pulse labeling experiment (Figs. 8.2–8.4) for the structural characterization of folding intermediates proceeds as follows (Roder et al., 1988; Englander and Mayne, 1992; Elöve et al., 1994; Dyson and Wright, 1996): Initially, the protein is unfolded in a denaturant–D$_2$O solution

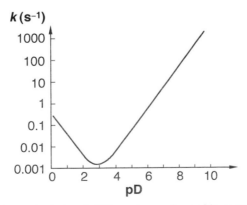

Fig. 8.1. Kinetic rate constant, k, of H/D exchange for amide protons of random chain polyalanine as function of pD (Englander and Mayne, 1992).

where exchangeable NH sites become deuterated. Then the unfolded protein is allowed to partially refold by mixing it with a deuterated buffer that favors refolding. After a variable time, the partially refolded protein is exposed to a H_2O labeling pulse of 5–50 ms duration at pH 9 where exchange of free hydrogens takes about 1 ms, so that amides at the unstructured parts of the protein become fully protonated but sufficiently protected sites stay deuterated. Then the pH is lowered to provide slow-exchange conditions and the protein is allowed to refold completely. Prior to the NMR measurement, the protein solution is concentrated by ultrafiltration.

Using the individual proton occupancies as function of the refolding time and the assignment of the resonances, the protected parts of early folding intermediates are mapped out. Kinetic information on a time scale of about 1 ms and longer can be obtained. Slightly higher time resolution, $\approx$200 µs, can be achieved for proteins which are stable at high pH using the double-jump mixing head illustrated in Fig. 5.5 (Sect. 5.1) and a labeling buffer at about pH 10.

H/D exchange kinetics is the currently available method with the highest structural resolution for monitoring early folding intermediates (Bycroft, 1990). Unfortunately, some proteins are prone to aggregation at low or high pH. Great care has to be taken not to confuse folding intermediates with aggregation.

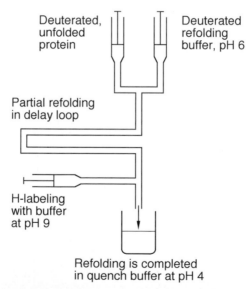

Fig. 8.2. Double-jump mixing for H/D exchange kinetic experiments (quenched-flow). H/D exchange of solvent-exposed amides is both acid and base catalyzed (see Fig. 8.1). The protein was stored in D_2O with a high denaturant concentration so that all exchangeable protons became replaced by deuterons. Refolding is initiated in the first mixer. After a certain period of time, determined by the velocity of the solution and the length of the delay loop, the partially refolded protein is H-labeled with buffer at pH about 9. Refolding of the labeled protein is completed in quench buffer at low pH where exchange is very slow.

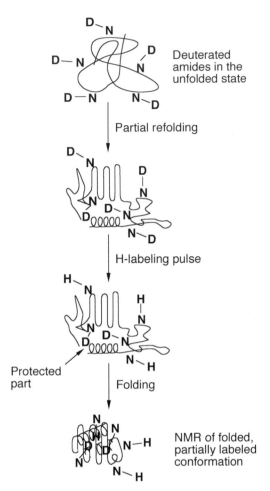

Fig. 8.3. Kinetic H/D exchange experiment. Prior to the refolding reaction, all exchangeable protons were replaced by deuterons in the unfolded state. Refolding is initiated with deuterated buffer. After partial refolding, the protein solution is mixed with H_2O. Parts of the protein which are protected at this time stay deuterated. Then H/D exchange is quenched with buffer of low pH and refolding is completed prior to NMR measurement.

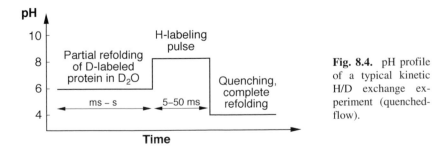

Fig. 8.4. pH profile of a typical kinetic H/D exchange experiment (quenched-flow).

8.2
Time-resolved circular dichroism

The kinetic resolution of single-wavelength detection of commercial CD spectrometers in combination with stopped-flow accessories is typically only a few
milliseconds, and the measurement of complete spectra is even slower.

Complete CD spectra of up to about 100 nm width may rapidly be recorded by
using a real-time multichannel spectrometer (Fig. 8.5; Nölting, 1991; Brandl et al.,
1991; Nölting et al., 1992). In this spectrometer, a polychromator instead of the
monochromator in scanning spectrometers (Fig. 7.9), and a multichannel detector
instead of a photomultiplier are used. A charged-coupled device (CCD) that has a
quantum efficiency of 60% at 600 nm wavelength, and still 20% at 200 nm,
serves as a detector with 512 channels.

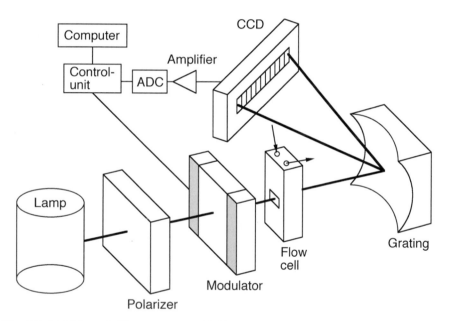

Fig. 8.5. Multichannel CD spectrometer (Nölting et al., 1992). The light of a deuterium- or
xenon lamp is first linearly polarized by the polarizer, and then circularly polarized by the
modulator. The modulator generates alternately left and right circularly polarized light. After
passing through the flow cell, the different wavelengths are separated by a holographic grating.
The spectrum is recorded by a charged-coupled device (CCD). The operation of the CCD is
synchronized with that of the modulator, so that in every cycle of the CCD the transmitted
intensity for only one of the two polarizations, left- or right-circular, is recorded. CD amplitudes
are calculated for every channel from the intensities of transmitted left and right circularly
polarized light. The absorption of an optimal signal-to-noise ratio for this measurement principle
is about 0.8–0.9, and sharply falls off with increasing absorption above 1.5 (see Fig. 8.7).

In the same way as for most scanning CD spectrometers (Sect. 7.2), the difference in absorption between left and right circularly polarized light, $\Delta A = A_1 - A_r$, is calculated using Eq. 8.1:

$$\Delta A \approx \frac{2 \log(e)}{f} \frac{I_1 - I_r}{I_1 + I_r} , \qquad (8.1)$$

where I_1 and I_r are the intensities of left and right circularly polarized light, respectively, $2 \log(e) \approx 0.87$, and f (here ≈ 0.9) is a correction factor for the degree of circular polarization. For a typical measurement on proteins, $\Delta A \approx 10^{-5} - 10^{-3}$.

Because all wavelengths are recorded simultaneously within a certain wavelength region, $\Delta\lambda$, for a given photon shot noise, the speed for recording a spectrum is increased by a factor of f_Δ:

$$f_\Delta = \Delta\lambda/\delta\lambda , \qquad (8.2)$$

where $\delta\lambda$ is the spectral resolution. For example, at a spectral resolution of 1 nm, a wavelength region of 100 nm may be measured 100 times faster than with a scanning CD spectrometer, at a given photon shot noise. Further, since the instrument does not need moving parts for the change of wavelength within a certain wavelength region, its operation is very stable and its construction is very compact. A short length of the optical path in this spectrometer is of advantage at short wavelengths where CD instruments have to be purged with nitrogen to remove absorbing oxygen, ozone, and water vapor.

For a signal, S, of photons converted into electrons per time unit by the CCD, the noise, σ_S, is approximately (Nölting, 1991; Stark et al., 1992)

$$\sigma_S^2 = S + \sigma_R^2 , \qquad (8.3)$$

where σ_R is the signal-independent noise of the detector (Fig. 8.6).

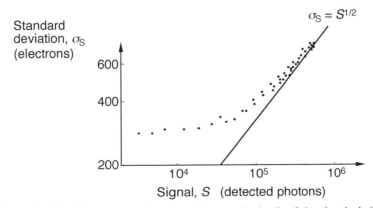

Fig. 8.6. Standard deviation, σ_S, as function of the magnitude, S, of the electrical signal per pixel for the CCD detector used for the multichannel CD spectrometer (Fig. 8.5). Only at low amplitudes is the signal dominated by the signal-independent noise. For this CCD, at signals $>10^5$ photons per pixel, the measurement is photon-shot-noise-limited, i.e., the noise is about equal to the square root of the number of detected photons (Nölting, 1991; Stark et al., 1992).

Thus, a good signal-to-noise ratio in fast kinetic measurements requires a large light throughput. That is why the arrangement of the optics of the multichannel CD spectrometer is specifically optimized for a very high light throughput: The light is focused onto the CCD device by a holographic grating with an aperture angle of 16°. The small number of optical elements and the high aperture of the grating ensures that for a non-absorbing sample about 50% of the light emitted by the lamp is transmitted to the CCD detector.

It should be pointed out, that in CD spectrometers the sensitivity of the measurement strongly depends also on the sample absorption (Fig. 8.7). Obviously, a small concentration of the sample causes a small CD amplitude and thus a low signal-to-noise ratio. On the other hand, a very large concentration causes a large CD amplitude, but also a large photon shot noise since the sample absorbs too much light. The optimal signal-to-noise ratio of a photon-shot-noise-limited CD measurement is achieved at an absorption of 0.87. Furthermore, at a very high absorption, CD spectrometers show a non-linear response. Thus, when measuring a CD spectrum, the sample should be adjusted to an extinction not outside the range of 0.1–1.5 for the whole wavelength region measured, if possible.

The polarization modulator (Fig. 8.8) acts as a retardation element and the modulation cycles convert linearly polarized light into alternately left and right circularly polarized light (Velluz et al., 1965; Nölting, 1991). In contrast to the sinusoidal modulation in commercial CD spectrometers, in this instrument the electro-optical modulator is operated with a nearly rectangular voltage profile (Fig. 8.9). This causes a smaller deviation of the polarization state of the light from the exactly circularly polarized state and thus, a further improvement in the signal-to-noise ratio. However, the suppression of artifacts caused by optical rotation is more complicated than in the classic design. The electro-optical modulator shown in Fig. 8.8 is transparent down to about 190 nm. For shorter wavelengths, photoelastic modulators are used.

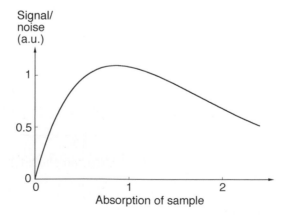

Fig. 8.7. Signal-to-noise ratio of a photon-shot-noise-limited CD measurement as function of the absorption of the sample (Nölting, 1991).

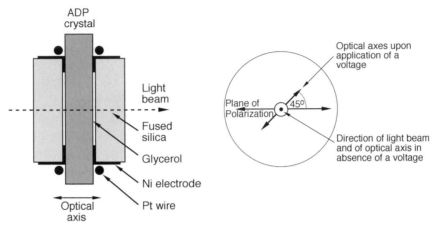

Fig. 8.8. Electro-optical modulator (Pockels cell) for the conversion of linearly into circularly polarized light (Nölting, 1991). Transparent electrodes are made by using glycerol which contains a few percent of water. The diameter of the z-cut ammonium dihydrogen phosphate (ADP) crystal is 40 mm, its thickness is 2 mm, and that of the glycerol layer is about 0.01 mm. In the absence of a voltage, the optical axis of the ADP crystal is in line to the incident linearly polarized light beam. Upon application of a voltage, the optical axis splits up into two axes which are tilted relative to the direction of the incident beam (see right side of the figure). The optical effect of these two axes is approximately the same as the effect of two optical axes, of which one is orientated parallel, and the other perpendicular, to the direction of the incident light beam. Then, the axis that is perpendicular to the incident beam has an orientation of 45° relative to the plane of linear polarization of the incident beam (see Fig. 7.10).

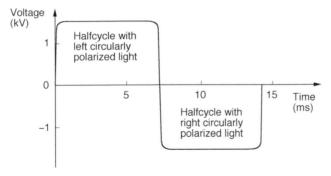

Fig. 8.9. Nearly rectangular voltage profile applied to the electro-optical modulator (Fig. 8.8) of the multichannel CD spectrometer (Fig. 8.5).

The time resolution of the real-time multichannel CD spectrometer (Fig. 8.5) is limited by the frequency of the read-out of the CCD detector and by the photon shot noise. Unfortunately, even the use of a significantly stronger light source would not enable a high-precision real-time measurement in the µs- or ns-time scale because available CCD devices and diode arrays can convert only relatively low photon currents.

Further extension of the time resolution is obtained by repetitive methods: The reaction is repeated many times and the signals are accumulated (Xie and Simon, 1989, 1991). Fig. 8.10 shows a repetitive single wavelength CD spectrometer with a 50-ps time resolution, limited by the lengths of the LASER pulses. To obtain one time point for one wavelength, the reaction is carried out several times for the left and right circular polarizations and the signals are averaged. An optical delay line serves for a time delay between pump and probe beam: In air, light travels 30 cm in 1 ns.

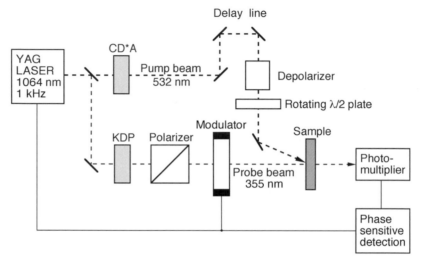

Fig. 8.10. Repetitive pump and probe beam CD spectroscopy with a time resolution of 50 ps. A pump beam is generated by frequency-doubling the 1064-nm output of a YAG LASER using a CD*A crystal. After careful depolarization, carried out with the help of depolarizer and rotating half-wavelength plate, the pump beam is passed onto the sample where it initiates the desired reaction kinetics. Relaxation in the sample is monitored at 355 nm with the help of the probe beam which is generated by a frequency-tripling stage using a KDP crystal. The probe beam is linearly polarized with a polarizer and polarization modulated with a photoelastic modulator that is timed to the LASER so that successive light pulses alternately pass through the modulator when it is in the $+\lambda/4$ and $-\lambda/4$ position. After passing through the sample, the transmitted probe beam is phase-sensitively detected (Xie and Simon, 1989, 1991).

A solution to the problem of the saturation of the detector at high photon currents has been shown by Kliger et al. (Lewis et al., 1985, 1992; Björling et al., 1996; Chen et al., 1998). The groundbreaking measurement principle (Figs. 8.11–8.13) enables CD spectroscopy in the time scale from picoseconds to seconds (see also Sect. 5.3.1). In this principle, the difference of intensity between transmitted left and right polarized light is optically amplified prior to the conversion into an electrical signal.

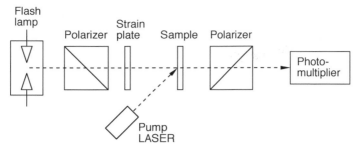

Fig. 8.11. CD measurement in the nanosecond time scale (Einterz et al., 1985; Lewis et al., 1985, 1987, 1992; Milder et al., 1988). Linearly polarized light generated by flash lamp and first polarizer is converted with the help of a strain plate into alternately left and right elliptically polarized light. In the absence of an optically active sample, a second polarizer absorbs most of the elliptically polarized light. However, placing a CD-active sample into the beam causes a significant change of the transmitted intensities (see Fig. 8.12). Compared to commercial spectrometers (Fig. 7.9), the relative difference of intensities is amplified by a factor of about $2.5\ \delta^{-1}$, where δ is the phase difference of the strain plate. In order to avoid artifacts, δ usually should not be smaller than 0.01 (0.6°), corresponding to an amplification by a factor of ≈ 250.

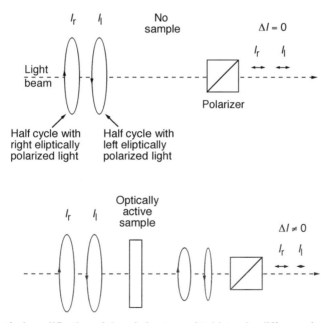

Fig. 8.12. Optical amplification of the relative transmitted intensity difference in a new type of fast CD spectrometers (Einterz et al., 1985; Lewis et al., 1985, 1987, 1992; Milder et al., 1988; Wen et al., 1996). Highly eccentric elliptically instead of circularly polarized light is used. A polarizer (second polarizer in Fig. 8.11) which is orientated perpendicular to the major axis of elliptical polarization absorbs most of the light in the absence of an optically active sample (top). In the presence of an optically active sample, the eccentricity of the light is changed which gives rise to a significant relative difference of the two light intensities after passing through the polarizer (bottom).

In contrast to the common design variants (Figs. 7.9 and 8.5), in this design, elliptically polarized light of high eccentricity is used to measure CD, and here the relative difference of detected light intensities is

$$\frac{I_l - I_r}{I_l + I_r} \approx \frac{1}{\log(e)} \frac{\Delta A}{\delta} \ , \tag{8.4}$$

where I_l and I_r are the transmitted intensities for left and right elliptical polarization, respectively, ΔA is the difference of absorptions between left and right circularly polarized light, and δ is the retardation of the strain plate (Figs. 8.11 and 8.12; Esquerra et al., 1997). In this way, a certain CD signal, ΔA, and also optical rotatory dispersion (Shapiro et al., 1995), may be detected with an about 100 times larger relative difference of the light intensity falling onto the electrical detector, corresponding to a 10.000 times faster detector-saturation- and photon-shot-noise-limited measurement. Also this measurement principle has been combined with multichannel detection (Fig. 8.13).

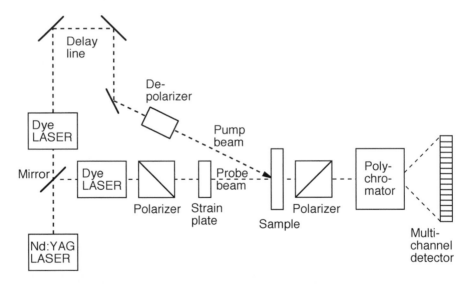

Fig. 8.13. Nanosecond multichannel CD spectroscopy. Pump (excitation) beam and probe beam are obtained from two dye LASERs which are synchronously pumped by the second harmonic of a mode-locked Nd:YAG LASER. Transients are obtained by scanning the delay line and taking several shots at each step. A change in the difference between the two optical paths in air by 0.3 mm corresponds to a 1-ps change of the delay between pump and probe beam. Optical activity in the sample is probed by using alternately left and right elliptically polarized light of high eccentricity that is generated with the first polarizer and a strain plate. A second polarizer which is orientated perpendicular to the first polarizer serves for a highly sensitive analysis of the polarization state of the probe beam after passing through the sample. Artifacts due to transient birefringence caused by the pump beam are diminished with the help of a depolarizer (Einterz et al., 1985; Lewis et al., 1985, 1987, 1992; Milder et al., 1988; Wen et al., 1996).

8.3
Φ-value analysis

All of the aforementioned kinetic methods in which rate constants of folding are obtained may be combined with a protein engineering approach (Goldenberg et al., 1989; Matouschek et al., 1989, 1990; Matouschek and Fersht, 1991, 1992, 1993, 1995a, b; Fersht et al., 1991, 1992; Otzen et al., 1994; Itzhaki et al., 1995b, Nölting et al., 1995, 1997a; Nölting and Andert, 2000; Nölting, 1998a, 1999, 2003). This method, the so-called Φ-value analysis, uses the build-up of inter-action energies in the protein molecule along the folding pathway as a measure of structure formation, and is of paramount importance for the high structural resolution of folding transition states and early folding intermediates (Fig. 8.14).

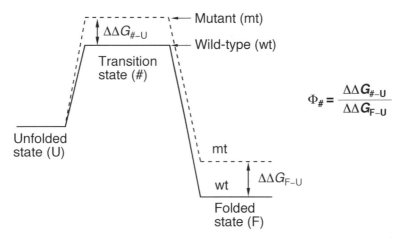

Fig. 8.14. Φ-value analysis. The case of a protein that folds via a two-state mechanism is illustrated. A mutation causes a change of stability, $\Delta\Delta G_{F-U}$, of the folded state (F) compared with the unfolded state (U). $\Delta\Delta G_{\#-U}$ is the energy difference between mutant and wild-type in the transition state, #. The fraction of energy difference, $\Phi_{\#} = \Delta\Delta G_{\#-U}/\Delta\Delta G_{F-U}$, depends on the amount of structure that has built up in # at the position of the mutation. $\Phi = 0$ corresponds to no structure formation; for complete formation of structure at the position of the mutation, $\Phi = 1$. A set of single mutants, strategically distributed over the molecule is used to map out the structure of the transition state at the resolution of single amino acid residues. For the assumptions made for this analysis see (Fersht et al., 1992, 1994). This approach may analogously be applied on proteins with more complicated energy landscapes which contain intermediates on the reaction pathway, and on proteins which have residual structure in their unfolded states.

The basic idea is to create a mutation in the protein molecule which causes a difference in stability between mutant and wild-type, $\Delta\Delta G_{F-U}$. This difference in stability builds up in the course of the folding reaction (Figs. 8.14, 8.15). Early in the folding reaction, when no structure has yet formed at the position of the mutation, there is no difference in Gibbs free energy, $\Delta\Delta G$, between mutant and wild-type. When, in the course of the folding reaction, structure has completely formed at the position of the mutation, the difference in stability between wild-type and mutant is as large as in the folded state, i.e., $\Delta\Delta G = \Delta\Delta G_{F-U}$. Using the energy as a structurally sensitive probe, and a set of non-disruptive mutants, strategically distributed over the molecule, the structures of folding intermediates can be mapped out for the same time scale for which kinetic methods are available.

Most importantly, this is the only existent method which can be used to determine the structures of transition states at the resolution of individual amino acid residues. The knowledge of the transition state structures is of paramount importance for understanding the protein folding problem since transition states represent, per definition, the highest points in the energy landscape along the folding pathway (see Sect. 4.1). Knowledge of these structures can give insight into the processes that determine the speed of folding and may be used to re-engineer the speed of folding in a rational way.

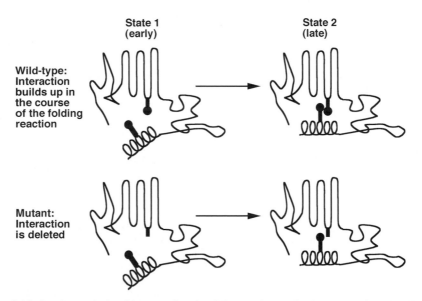

Fig. 8.15. Φ-value analysis with a mutation that deletes an interaction in the protein molecule: In the course of the folding reaction of wild-type protein, an interaction builds up which causes a change of energy. In the mutant, this interaction is deleted. By comparing the energies of mutant and wild-type protein in all individual steps of the folding reaction one can determine at which stage of the folding reaction this interaction has formed. If at a certain (early) time point the interaction is not formed yet, there is no difference in energy (and kinetics) between mutant and wild-type: an interaction which does not exist, makes no physical difference. If, in contrast structure is completely formed at the position of the mutation at a later time point, then at this time point the difference in stability between mutant and wild-type is as large as in the folded state. Thus, the deletion of interactions serves as a probe of structure consolidation; the Gibbs free energy difference is used as a measure of the degree of structure consolidation (see also Figs. 8.14, 8.29, and 8.30). A set of mutants serves for mapping out the structural changes along the folding pathway throughout the whole molecule.

8.3.1
Protein engineering

The Φ-value analysis requires the availability of a set of non-disruptive mutants, i.e., mutants in which the structure is not significantly altered outside the position of the mutation (Matouschek et al., 1989, 1990; Matouschek and Fersht, 1991). In hydrophobic cores of the protein molecule, preferentially the mutations delete only one or two methyl-groups, for example Val→Ala, Leu→Ala, Ile→Ala, Ala→Gly. In helices mutations may be used which delete interactions of neighboring amino acid residues, for example Gln→Ala, Gln→Gly, Ala→Gly, and Ser→Ala.

There is an ongoing progress in the development of simpler and faster methods for protein engineering (see, e.g., Ausubel et al., 1992). For reasons of brevity, here only the principles of two very robust methods of site-directed mutagenesis are explained, the PCR-mediated mutagenesis (PCR = polymerase chain reaction), and the cassette mutagenesis. Currently, the net-time effort for preparing, expressing and purifying a set of 60 point mutants may be less than 2 months for one person when using PCR-mediated mutagenesis.

The initial step for protein genetic engineering is to clone the wild-type protein. Usually a plasmid, which is a circular piece of DNA (deoxyribonucleic acid) of typically a few 1000 base pairs in length (Fig. 8.16), is used as a cloning vector. The gene of the protein can be obtained by isolation from the host or by synthesis. Insertion of the gene into the vector may similarly be done as in cassette mutagenesis (Sect. 8.3.1.1). A plasmid which contains the DNA that encodes for the protein is called a clone. Clones for numerous proteins may be obtained from several laboratories of large-scale sequencing projects. For clones which are published in the literature there is usually an honorary obligation for the authors to make these available to other laboratories for non-commercial research purposes.

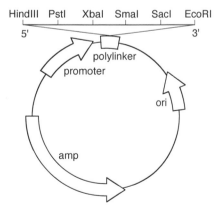

Fig. 8.16. Example for a plasmid cloning vector with a promoter and the origin of replication (ori). DNA which encodes for a protein (the gene) may be inserted into the polylinker site that contains 6 restriction sites in this example. In order to simplify the selection of transformants, this plasmid is engineered to carry the gene for antibiotic resistance (amp).

Four steps are necessary for the site-directed mutagenesis presented here (Ausubel et al., 1992; Creighton, 1993):

1. Isolate the plasmid containing the gene of wild-type protein from the host cells via a miniprep.
2. Mutagenesis reactions (see Sects. 8.3.1.1 and 8.3.1.2).
3. Insert the mutant plasmid into the host cells and let the transformed cells grow.
4. Isolate the plasmid with the mutant gene from the host cells via a miniprep, and sequence the gene to verify the mutation.

8.3.1.1
Cassette mutagenesis

A fragment of the DNA encoding for the wild-type protein, a so-called cassette, is cut out with the help of specific restriction enzymes at the unique restriction sites A and B (Fig. 8.17). The cloning vector obtained is then thoroughly purified by gel electrophoresis. Two mutagenic synthetic oligonucleotides are hybridized to form a mutagenic cassette which is inserted into the cloning vector between points A and B.

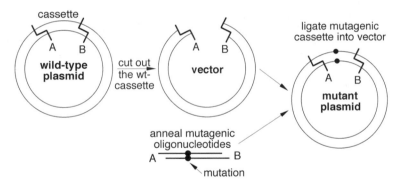

Fig. 8.17. Cassette mutagenesis of plasmid DNA (Ausubel et al., 1992).

8.3.1.2
PCR mutagenesis

The availability of restriction sites limits the applicability of the cassette mutagenesis. Since the purity of synthetic oligonucleotides rapidly decreases with increasing length, it is impractical to use the cassette mutagenesis if there are no suitable restriction sites nearby the position of the intended mutation. With the technique of polymerase chain reaction (PCR), DNA of many 100 base pairs (BP) length may easily be synthesized with less than 1 error per 10,000 BP on average.

In the PCR reaction (Figs. 8.18, 8.19), the target DNA is amplified in a cyclic reaction containing three simple steps:

1. Specific hybridization of two oligonucleotides (so-called primers) with complementary, or nearly complementary, sequences on the template DNA.
2. Extension of the primers by a polymerase.
3. Thermal denaturation, i.e., dissociation, of the double stranded DNA, and return to step 1.

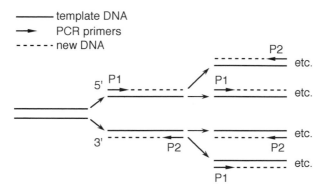

Fig. 8.18. Polymerase chain reaction (PCR). At higher temperature, the double stranded template DNA dissociates. After re-cooling, single stranded DNA may bind specifically to primers that have a complementary or nearly complementary sequence. At high concentrations of primers, compared with template DNA, the binding of template DNA to primers is favored over the unproductive binding of two strands of template DNA. Now the polymerase may extend the DNA (dashed lines). The procedure of (a) denaturation, (b) binding of the primers, and (c) extension is repeated, typically 20 to 40 times. This can result in a many million-fold amplification of a specific DNA fragment. Efficiencies per cycle can exceed 70% (see also Sect. 9.1.2).

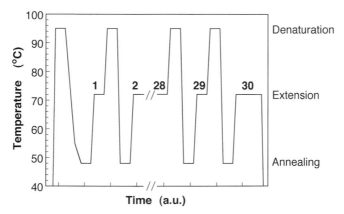

Fig. 8.19. Example for the temperature profile in a PCR with 30 cycles. During the extension reaction, a polymerase catalyzes the growth of new DNA strands in a 5'→3' direction. Extension time and temperature depend on the type of polymerase used and on the length of the DNA fragment that is to be amplified. For example, for *Thermus aquaticus* DNA polymerase, the optimal temperature for polymerization is about 70–80°C, and the rate of nucleotide incorporation under optimal reaction conditions may exceed 40 nucleotides per second.

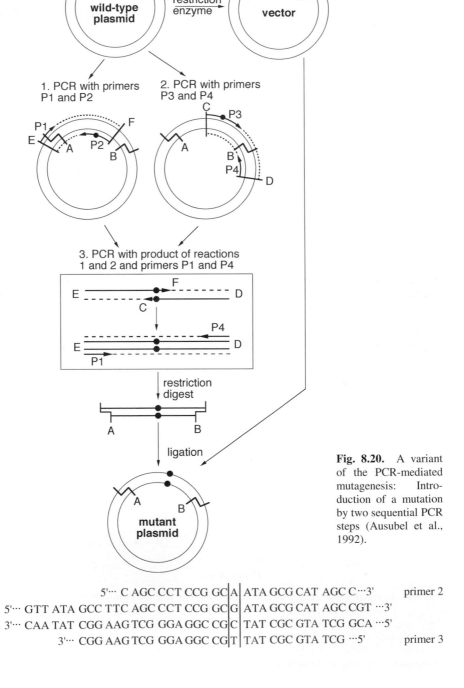

Fig. 8.20. A variant of the PCR-mediated mutagenesis: Introduction of a mutation by two sequential PCR steps (Ausubel et al., 1992).

```
                5'··· C AGC CCT CCG GC A ATA GCG CAT AGC C···3'              primer 2
  5'··· GTT ATA GCC TTC AGC CCT CCG GC G ATA GCG CAT AGC CGT ···3'
  3'··· CAA TAT CGG AAG TCG GGA GGC CG C TAT CGC GTA TCG GCA ···5'
        3'··· CGG AAG TCG GGA GGC CG T TAT CGC GTA TCG ···5'                 primer 3
```

In the variant of the PCR-mediated mutagenesis illustrated in Fig. 8.20, the whole gene of a protein is amplified (Ausubel et al., 1992):

1. Two overlapping fragments of the mutant gene for the protein including flanking regions are prepared in two PCR reactions.
2. The product of these two reactions serves as the template for the third PCR in which the whole mutant gene is combined and amplified.
3. The product of the third PCR is digested at the positions A and B which are located outside the gene.
4. For the preparation of the cloning vector, the wild-type (wt) gene is removed from the wt-plasmid with the help of specific restriction enzymes that digest the plasmid specifically at two unique restriction sites, A and B. The cloning vector obtained is thoroughly purified and any undigested DNA removed to avoid wt-contamination in the later steps.
5. The purified digest of the third PCR is ligated into the vector.

8.3.2
Determination of the protein stability in equilibrium

A convenient method to determine the stability of a protein is the equilibrium titration with denaturants: The protein is unfolded with denaturants and a spectroscopic signal, for example, fluorescence or circular dichroism, is recorded as function of the concentration of denaturant (see, e.g., Privalov, 1979; Matouschek et al., 1989). From the curves obtained with sigmoidal shape (Fig. 8.21), one can calculate the Gibbs free energy change upon folding of the protein, ΔG_{F-U}, as described in the following paragraphs.

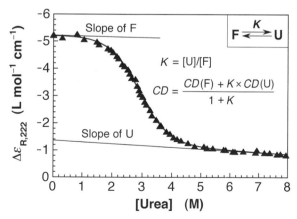

Fig. 8.21. Urea-induced unfolding of P27A/C40A/C82A barstar at 25°C, monitored by the mean residual ellipticity at 222 nm, $\Delta \varepsilon_{R,222}$, and fitted to Eq. 8.13 with β_F set to 0. K is the equilibrium constant for unfolding ($K = [U]/[F]$), that is sometimes also marked as "K_{-1}".

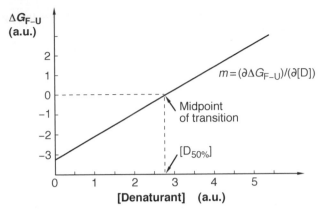

Fig. 8.22. Linear change of the stability of proteins with the concentration of denaturant. The Gibbs free energy change upon folding, ΔG_{F-U}, is negative under conditions that favor folding.

Usually ΔG_{F-U} increases, in good approximation, linearly with the denaturant concentration, [D] (Fig. 8.22; Matouschek et al., 1989):

$$\Delta G_{F-U}([D]) = \Delta G_{F-U}(0) + m \times [D] , \qquad (8.5)$$

where $m = \partial(\Delta G_{F-U}([D]))/\partial[D]$. This is because denaturants, for example, urea and guanidine hydrochloride (GuHCl), are better solvents for the unfolded state than for the folded state. Urea and GuHCl are supposed to form hydrogen bonds to the polar groups of the polypeptide backbone, and thereby, to favor the more solvent-exposed state, i.e., the unfolded state.

Note that for a stable protein in water ΔG_{F-U} has a negative sign. According to Eq. 8.5, ΔG_{F-U} in the absence of denaturants is

$$\Delta G_{F-U}(0) = -m \times [D]_{50\%} , \qquad (8.6)$$

where $[D]_{50\%}$ is the denaturant concentration at the midpoint of the equilibrium between folded and unfolded state, i.e., $\Delta G_{F-U}([D]_{50\%}) = 0$. Thus,

$$\Delta G_{F-U}([D]) = m \times ([D] - [D]_{50\%}) . \qquad (8.7)$$

Consider a two-state transition,

$$U \underset{}{\overset{K}{\rightleftharpoons}} F , \qquad (8.8)$$

where K is the equilibrium constant for unfolding, i.e., the ratio of the population of the unfolded state, [U], relative to that of the folded state, [F],

$$K = [U]/[F] . \qquad (8.9)$$

The observed signal, S, is given by the sum of the signal of the folded state times its population, plus the signal of the unfolded state times its population:

$$S = \{\alpha_F + \beta_F[D] + (\alpha_U + \beta_U[D])K\}/(1+K), \tag{8.10}$$

where α_F, β_F, α_U, and β_U define linear baselines for the signals of the folded and unfolded states as function of denaturant concentration, respectively. The equilibrium constant for unfolding, K, is related with the Gibbs free energy change upon folding, ΔG_{F-U}, by Eq. 8.11:

$$K = \exp\{\Delta G_{F-U}/(RT)\}, \tag{8.11}$$

where R and T are the universal gas constant and absolute temperature, respectively. According to Eqs. 8.7 and 8.11, K may be expressed as follows:

$$K = \exp\{m([D] - [D]_{50\%})/(RT)\}. \tag{8.12}$$

Now Eq. 8.12 may be inserted into Eq. 8.10:

$$S = \frac{\alpha_F + \beta_F[D] + (\alpha_U + \beta_U[D])\exp\{m([D]-[D]_{50\%})/(RT)\}}{1 + \exp\{m([D]-[D]_{50\%})/(RT)\}}. \tag{8.13}$$

The unfolding curve, e.g., in Fig. 8.21, is fitted to Eq. 8.13, and so, m and $[D]_{50\%}$ are obtained which determine the change of Gibbs free energy upon folding in the absence of denaturants according to Eq. 8.6 ($\Delta G_{F-U}(0) = -m \times [D]_{50\%}$).

For CD detection, occasionally, β_F is very small and may be set to zero without significantly altering the calculated $\Delta G_{F-U}(0)$. This may result in a better stability of the curve fit (see Sect. 9.4).

Eqs. 8.5–8.13 are derived for two-state transitions. However, Eqs. 8.5–8.13 may also be applied on a multi-state transition involving a number of kinetic intermediates if the population of the intermediates in equilibrium is small. Further, small systematic errors usually cancel out to a large degree when comparing only differences of Gibbs free energies between different mutants as done in the Φ-value analysis.

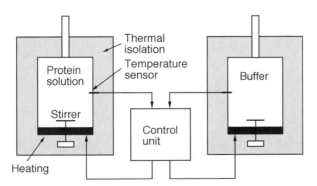

Fig. 8.23. Principle of differential scanning calorimetry (DSC): Two identical sample cells, one containing the protein solution, and the other the buffer, are heated at the same constant rate, typically 0.1–1 K min^{-1}. The difference of the heat uptake between both cells is measured by monitoring the electrical energy needed for the temperature increase.

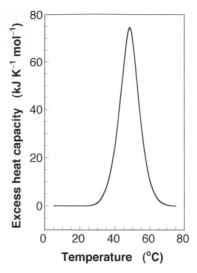

Fig. 8.24. Differential calorimetric scan of cytochrome P-450cam (Pfeil et al., 1993a).

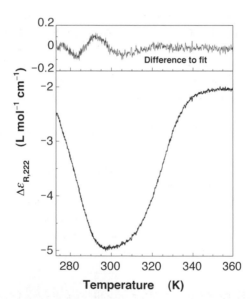

Fig. 8.25. CD T-scan of C40A/C82A/P27A bar-star in 2 M urea (bottom), and the difference to a curve fit with Eq. 8.14 (top).

Alternatively, ΔG_{F-U} may be measured using calorimetry, e.g., (Privalov, 1979; Pfeil, 1981, 1993; Pfeil et al., 1993a, b). In differential scanning calorimetry, the protein is refolded or unfolded by changing the temperature, and the heat emission or heat absorption during the reaction is measured (Figs. 8.23, 8.24). Relatively large quantities of protein are needed for calorimetry, typically 1–10 mg protein dissolved in 1 mL buffer, but enthalpy and heat capacity are also obtained.

The third common method for the determination of ΔG_{F-U} is to use thermal scans, i.e., to measure a spectroscopic signal of the protein in the transition region as function of temperature (e.g., Fig. 8.25). Usually the curve is fitted to

$$Y(T) = \frac{Y_F + m_F T + (Y_U + m_U T)K}{1 + K} \qquad (8.14)$$

$$K = \exp(\frac{\Delta H_{F-U}(1 - T/T_g) + \Delta C_p(T - T_g - T\ln(T/T_g))}{RT}) \; ,$$

where T is the absolute temperature, R the universal gas constant, T_g the temperature of half transition, $Y(T)$ is the signal at T, Y_F and Y_U are the signals of folded and unfolded state at $T = 0$, respectively, m_F and m_U set the slopes of the signals of the folded and unfolded state, respectively, ΔH_{F-U} is the enthalpy change upon folding at $T = T_g$, and ΔC_p the heat capacity change at $T = T_g$. Then the Gibbs free energy change upon folding is simply given by

$$\Delta G_{F-U} = RT \ln (K) . \qquad (8.15)$$

However, often the quality of the fit is unsatisfactory (Fig. 8.25), partially because of aggregation at higher temperature. Even more importantly, Eq. 8.14 contains 7 free parameters which are fitted to a curve of relatively simple shape, and so the fit is often unstable and may lead to wrong results. Sometimes, improved results are obtained with Eq. 8.16 which takes into account that the baseline of the unfolded state is not a straight line, but is slightly curved:

$$Y(T) = \frac{Y_F + m_F(T - T_o) + (Y_\infty + (Y_U - Y_\infty)\exp(m_U(T - T_o)))K}{1 + K} \; , \qquad (8.16)$$

where T_o is a conveniently chosen fixed temperature, and here Y_F and Y_U are the signals of folded and unfolded state at T_o, respectively, and Y_∞ is the signal of the unfolded state at infinite temperature in the case of $m_U < 0$. m_U and the, often small, m_F may be determined outside the transition region and inserted into Eq. 8.16. Still, this method is only advisable for proteins with little if any aggregation, and when a very wide range of temperature is accessible.

8.3.3
Measurement of kinetic rate constants of folding and unfolding

Measurements of kinetic rate constants have to be done with care: The occurrence of transient aggregation has to be checked spectroscopically and by comparing the rate constants and relative amplitudes between different concentrations of protein (see Sects. 8.3.3.4 and 9.2). It should be noted that the concentration-independence of the observed rate constant is not sufficient to rule out a monomer–multimer mechanism since the observed kinetics might be dominated by the dissociation reaction which has a concentration-independent rate constant.

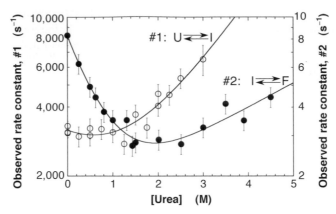

Fig. 8.26. Observed rate constants as function of the urea concentration for the two folding transitions, #1 and #2, of the 10 kDa protein barstar (see Sect. 10.5.2).

For the Φ-value analysis both the folding (k_1, k_2, ...) and unfolding (k_{-1}, k_{-2}, ...) rate constants have to be known. This information is extracted by decomposition of the curves of the observed rate constants (k_{1obs}, k_{2obs}, ...) as functions of the concentrations of denaturant and extrapolation to zero denaturant concentration. The observed rate constants display a V-shaped behavior where they are dominated by the folding rate constants at low denaturant concentrations, and by the unfolding rate constants at high denaturant concentrations (Fig. 8.26).

8.3.3.1
Two-state kinetics

For a two-state transition between the unfolded, U, and folded, F, state,

$$U \underset{k_{-1}}{\overset{k_1}{\rightleftharpoons}} F , \qquad (8.17)$$

the observed rate constant, k_{obs}, as a function of the concentration of denaturant, [D], is given by Eq. 8.18 (see Sect. 4.2):

$$k_{obs}([D]) = k_1(0)\, e^{-m_1[D]} + k_{-1}(0)\, e^{m_{-1}[D]} , \qquad (8.18)$$

where m_1 and m_{-1} determine the slopes of the rate constants for folding and unfolding, respectively, as a function of [D] (see Sect. 8.3.2). Sometimes, k_{obs} is also called the relaxation constant. In order to obtain $k_1(0)$ and $k_{-1}(0)$, Eq. 8.18 is fitted to the curve of $k_{obs}([D])$ (Nölting et al., 1995; Nölting et al., 1997a). For the Φ-value analysis, the ratios of $k_1(0)$ and $k_{-1}(0)$ between mutant and wild-type are needed. In the case of only small differences, if any, in m_1 and m_{-1} between mutant and wild-type, using the same m-values for mutant and wild-type may improve the stability of the fit, and small errors may partially cancel each other out.

8.3.3.2
Three-state kinetics

The general solutions for three-state transitions are given in Sect. 4.3. For a consecutive three-state transition, composed of a slow and a much faster transition,

$$U \underset{k_{-1}}{\overset{k_1}{\rightleftharpoons}} I \underset{k_{-2}}{\overset{k_2}{\rightleftharpoons}} F , \qquad (8.19)$$

where U, I, and F designate the unfolded, intermediate, and folded state, respectively, the observed slow rate constant, k_{2obs}, is approximately

$$k_{2obs} \approx k_2 \{K_{-2} + 1/(1 + K_{-1})\} = k_{-2} + k_2/(1 + K_{-1}) \qquad (8.20)$$

$$K_{-2} = k_{-2}/k_2$$

$$K_{-1} = k_{-1}/k_1 ,$$

where k_1 , k_{-1} are the rate constants of the fast folding transition, and k_2 , k_{-2} are the rate constants of the slow folding transition (Nölting et al., 1995, 1997a; see also Sect. 4.3). Analogous to the exact solution for a two-state transition, the observed fast rate constant k_{1obs} is approximately the sum of the folding and unfolding rate constants (Nölting et al., 1995, 1997a),

$$k_{1obs} \approx k_1 + k_{-1} = k_1 (1 + K_{-1}) . \qquad (8.21)$$

Thus, Eq. 8.18 applies analogously also for the fast transition of a three-state folding mechanism, but Eq. 8.20 has to be applied for the extrapolation of the slow rate constants to zero concentration of denaturant.

8.3.3.3
Kinetic implications of the occurrence of intermediates

According to Eq. 8.20, k_{2obs} is always smaller than it would be in the absence of an intermediate, namely $k_2 + k_{-2}$ (see Eq. 8.18), and thus, the occurrence of an early intermediate retards folding. This retardation of folding is often the source of the "roll-over" effect (Fig. 8.27) which is a deviation of the measured curve for the observed rate constant as function of the concentration of denaturant from the shape expected for a two-state transition. However, depending on the stabilities of the species involved as function of the denaturant concentration, the roll-over effect may be quite small and easily remain undetected. On the other hand, from the presence of the roll-over effect one cannot necessarily conclude the existence of a folding intermediate. Transient aggregation may cause a similar effect.

There are two interesting implications of the occurrence of unfolding intermediates, i.e., of intermediates that are located on the reaction coordinate between

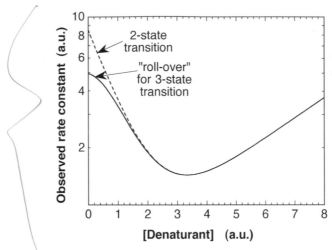

Fig. 8.27. Roll-over effect due to the presence of an early folding intermediate: At low denaturant concentration, the observed rate constant is slowed down relative to that expected for a two-state transition.

the folded and main transition states (see Sect. 5.2.1.3): First, the structure of the main transition state cannot be very close to the structure of the folded state, and second, similar to (re)folding intermediates, the observed rate constant is decelerated: For a three-state transition,

$$U \underset{k_{-1}}{\overset{k_1}{\rightleftharpoons}} I \underset{k_{-2}}{\overset{k_2}{\rightleftharpoons}} F , \qquad (8.22)$$

here composed of a slow transition between unfolded state, U, and intermediate state, I, and a much faster transition between I and folded state, F, the observed slow rate constant, k_{1obs}, is approximately

$$k_{1obs} \approx k_{-1} \{K_1 + 1 / (1 + K_2)\} = k_1 + k_{-1} / (1 + K_2) \qquad (8.23)$$
$$K_1 = k_1/k_{-1}$$
$$K_2 = k_2/k_{-2} ,$$

where k_1, k_{-1} are the rate constants of the slow transition, and k_2, k_{-2} are the rate constants of the fast transition (derived from Eq. 8.20). For a two-state transition, the observed slow rate constant is given by the sum of the folding and unfolding rate constants, $k_1 + k_{-1}$, which is always larger than the right-hand term in Eq. 8.23. Thus, both (re)folding and unfolding intermediates decrease the observed rate constant for the folding reaction; both retard folding (Nölting, 1996).

8.3.3.4
Discrimination between folding and association events

In experiments with small-amplitude T-jumps or small changes of denaturant concentration, the observed rate constant, k_{obs}, for a monomer–dimer equilibrium,

$$2A \underset{k_{-1}}{\overset{k_1}{\rightleftharpoons}} A_2 \qquad (8.24)$$

is

$$k_{obs} = 4[A] k_1 + k_{-1} , \qquad (8.25)$$

where $4[A]k_1$ and k_{-1} are the rate constants of association and dissociation, respectively (k_1 is the second-order rate constant of association).

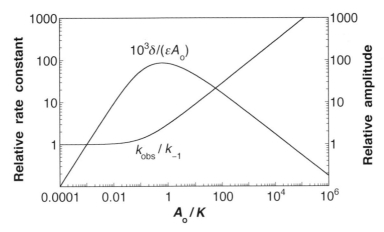

Fig. 8.28. Relative amplitude, $\delta/(\varepsilon A_o)$, ($\times 1000$) and relative observed rate constant, k_{obs}/k_{-1}, of the relaxation as function of the relative concentration, A_o/K, for T-jumping of a monomer–dimer equilibrium (Eq. 8.24), in which the forward and backward reactions have the same transition state. $K = k_{-1}/k_1$ and A_o is the total protein concentration. At low protein concentrations, corresponding to a small A_o/K, the observed rate constant is concentration-independent and the relative amplitude of the relaxation increases with increasing concentration. At high protein concentrations, corresponding to a large A_o/K, the observed rate constant increases and the relative amplitude decreases with increasing protein concentration (see also Sect. 4.6).

At high protein concentrations, i.e., under conditions where the major fraction in equilibrium is A_2, the observed rate constant is dominated by the concentration-dependent association rate constant (Fig. 8.28). At low concentrations, i.e., under conditions where A is the major fraction in equilibrium, the observed rate constant is dominated by the concentration-independent dissociation rate constant. In this case, the relative amplitude of the relaxation is concentration-dependent: A T-jump changes the constant $K = k_{-1}/k_1 = [A]^2/[A_2]$ approximately to $(1 + \varepsilon) K = ([A] + 2 \delta)^2/([A_2] - \delta)$, where ε and δ are small perturbations. Thus,

$$\delta \approx \varepsilon \, [A][A_2]/([A] + 4[A_2]) \approx \varepsilon \, [A_2] \; . \tag{8.26}$$

So, the relative amplitude is proportional to $[A_2]/A_o$, in which $A_o = [A] + 2\,[A_2] \approx [A]$. Doubling of A_o at low concentrations leads to an approximately 4–fold increase of $[A_2]$, and thus, to a doubling of the relative amplitude of the observed kinetics. Thus, dimerization or decay of dimers can be ruled out as the origin of a relaxation when both the rate constant and the relative amplitude do not change over a wide range of concentration. Also for more complicated reactions involving association/dissociation events, a concentration-dependence of the relative amplitude of relaxation is predicted, which may be weaker, however.

8.3.4
Calculation and interpretation of Φ-values

8.3.4.1
Two-state transition

For a two-state transition (Fig. 8.29; Fersht, 1985; Matouschek et al., 1989), the Gibbs free energy difference, $\Delta\Delta G_{\#-U}$, of the transition state (#) between mutant and wild-type is (the unfolded state is taken as the reference state):

$$\Delta\Delta G_{\#-U} = -RT \times \ln\left(\frac{k_{1,mt}}{k_{1,wt}}\right) \; , \tag{8.27}$$

where $k_{1,mt}$ and $k_{1,wt}$ are the rate constants for folding of mutant and wild-type, respectively (see Sect. 4.1).

In the Φ-value analysis, $\Delta\Delta G_{\#-U}$ is compared with the difference in the Gibbs free energy changes upon folding between mutant and wild-type $\Delta\Delta G_{F-U}$, i.e., with the difference in the stabilities of the folded states between mutant and wild-type,

$$\Delta\Delta G_{F-U} = \Delta G_{F-U,mt} - \Delta G_{F-U,wt} \tag{8.28}$$

$$\Phi_\# = \frac{\Delta\Delta G_{\#-U}}{\Delta\Delta G_{F-U}} \; , \tag{8.29}$$

where $\Delta G_{F-U,mt}$ and $\Delta G_{F-U,wt}$ are the changes in Gibbs free energy upon folding for mutant and wild-type, respectively.

Because

$$\Delta\Delta G_{F-U} = \Delta\Delta G_{\#-U} + \Delta\Delta G_{F-\#} \; , \tag{8.30}$$

analogously one can determine the Φ-value of # by using the rate constants for unfolding:

$$\Phi_\# = 1 - \frac{\Delta\Delta G_{F-\#}}{\Delta\Delta G_{F-U}} \tag{8.31}$$

$$\Delta\Delta G_{F-\#} = RT \times \ln\left(\frac{k_{-1,mt}}{k_{-1,wt}}\right) \; . \tag{8.32}$$

If the Φ-values, $\Phi_\#$, of the transition state determined by Eqs. 8.29 and 8.31 do not match, the protein under consideration does not fold via a two-state mechanism, provided artifacts of measurement are excluded.

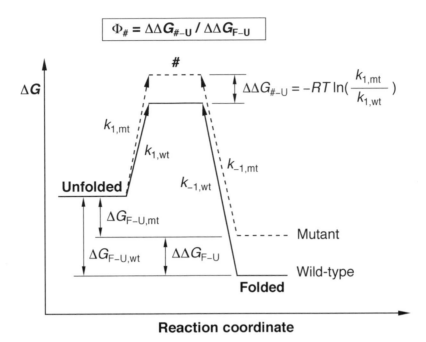

$$\Phi_\# = \Delta\Delta G_{\#-U} / \Delta\Delta G_{F-U}$$

Reaction coordinate

Fig. 8.29. Φ-value analysis for a two-state transition. In the folded state, mutant and wild-type protein differ in stability by $\Delta\Delta G_{F-U}$, which may be measured by equilibrium titration of the two proteins with denaturants (see Sect. 8.3.2). This energy difference builds up in the course of the folding reaction (the unfolded state is taken as the reference state). By using the kinetic rate constants for wild-type and mutant, one can find out whether an energy difference already exists in the transition state, #. The ratio, $\Phi_\#$, of the Gibbs free energy difference between mutant and wild-type in #, $\Delta\Delta G_{\#-U}$, relative to $\Delta\Delta G_{F-U}$ is taken as a measure of structure formation at the position of the mutation in # (see also Figs. 8.14 and 8.15).

8.3.4.2
Multi-state transition

For every step in the folding reaction which leads from the state i to the state $j=i+1$, the increase of the Φ-value is (Nölting et al., 1995, 1997a)

$$\Delta\Phi_{j-i} = \frac{\Delta\Delta G_{j-i}}{\Delta\Delta G_{F-U}} , \tag{8.33}$$

where $\Delta\Delta G_{j-i}$ is the difference in the Gibbs free energy changes for this reaction step between mutant (mt) and wild-type (wt). $\Delta\Delta G_{j-i}$ is calculated from the folding rate constants, $k_{i,\text{mt}}$ and $k_{i,\text{wt}}$, using Eq. 8.34 for a reaction step that is leading from a less folded state to a transition state, and from the unfolding rate constants, $k_{-i,\text{mt}}$ and $k_{-i,\text{wt}}$, using Eq. 8.35 for a reaction step leading from a transition state to a more folded state:

$$\Delta\Delta G_{j-i} = -RT \times \ln\left(\frac{k_{i,\text{mt}}}{k_{i,\text{wt}}}\right) \tag{8.34}$$

$$\Delta\Delta G_{j-i} = RT \times \ln\left(\frac{k_{-i,\text{mt}}}{k_{-i,\text{wt}}}\right) . \tag{8.35}$$

The Φ-value of the state n is given by summing the increments of the Φ-values,

$$\Phi_n = \sum_{j=2}^{n} \Delta\Phi_{j-i} , \tag{8.36}$$

where i denotes the state previous to the state j, and the unfolded state has the number 1.

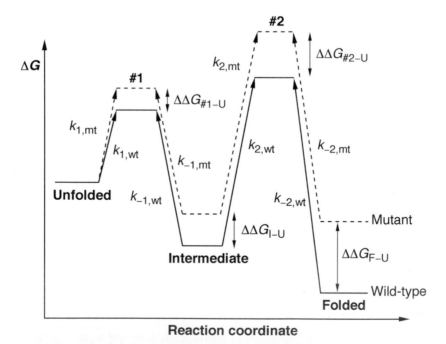

Reaction coordinate

Fig. 8.30. Φ-value analysis of a three-state transition. For each of the 4 steps of the reaction, $U \rightleftarrows \#1$, $\#1 \rightleftarrows I$, $I \rightleftarrows \#2$, and $\#2 \rightleftarrows F$, the kinetic rate constants are measured for mutant (mt) and wild-type (wt) protein. Then, using the kinetic rate constants and the total difference in Gibbs free energy changes upon folding between mutant and wild-type protein, $\Delta\Delta G_{F-U}$ (measured in equilibrium), the changes of Φ-value for every step are calculated. The total of the 4 increments of Φ-value must be 1 for all mutants if the protein folds according to the assumed on-pathway three-state mechanism (see text). A lower total of the Φ-value increments may be an indication for the occurrence of a further folding event of significant free energy contribution.

For example, Φ of the intermediate state, I, of a three-state folding reaction according to Fig. 8.30 is

$$\Phi_I = \frac{\Delta\Delta G_{\#1-U} + \Delta\Delta G_{I-\#1}}{\Delta\Delta G_{F-U}} = \frac{\Delta\Delta G_{I-U}}{\Delta\Delta G_{F-U}} \tag{8.37}$$

$$\Delta\Delta G_{\#1-U} = -RT \times \ln(\frac{k_{1,mt}}{k_{1,wt}}) \qquad \Delta\Delta G_{I-\#1} = RT \times \ln(\frac{k_{-1,mt}}{k_{-1,wt}}) \, ,$$

and for #2 we have

$$\Phi_{\#2} = \frac{\Delta\Delta G_{\#1-U} + \Delta\Delta G_{I-\#1} + \Delta\Delta G_{\#2-I}}{\Delta\Delta G_{F-U}} = \frac{\Delta\Delta G_{\#2-U}}{\Delta\Delta G_{F-U}} = 1 - \frac{\Delta\Delta G_{F-\#2}}{\Delta\Delta G_{F-U}} \tag{8.38}$$

$$\Delta\Delta G_{\#2-I} = -RT \times \ln(\frac{k_{2,mt}}{k_{2,wt}}) \qquad \Delta\Delta G_{F-\#2} = RT \times \ln(\frac{k_{-2,mt}}{k_{-2,wt}}) \, .$$

If the correct reaction mechanism was assumed, the Φ-value for the folded state, Φ_F, determined by Eq. 8.36 must be 1: The total of the energy differences obtained from the kinetic experiments ($\Delta\Delta G_{j-i}$) must be equal to the energy difference obtained from equilibrium measurements ($\Delta\Delta G_{F-U}$), unless a transition was left out in the analysis. Like other kinetic–thermodynamic methods, the Φ-value analysis cannot prove reaction mechanisms, but can only rule out alternatives. However, if for all mutants in a large set of mutants, the total of the Φ-value increments ($\Delta\Phi_{j-i}$) for the whole reaction ($=\Phi_F$) is 1, and the mutations are well distributed over the protein molecule, then it is very likely that no global folding event with significant energy contribution is missing in the analysis.

8.3.4.3
Residual structure in the unfolded state

In the previous sections the unfolded state was taken as an unstructured reference state. However, often unfolded states of proteins differ from the random coil state and may not be considered as fully unfolded (Nölting et al., 1997b). It can be shown that in case of residual structure in the (partly) unfolded state, the Φ-values calculated as shown in the previous sections represent the degree of formation of structure on top of the amount of residual structure in the unfolded state. In particular $\Phi = 0$ indicates the same degree of structure as in the unfolded state and $\Phi = 1$ is still obtained for complete structure consolidation in the state under consideration unless there are non-native interactions.

9 Experimental problems of the kinetic and structural resolution of reactions that involve proteins

9.1
Protein expression problems

9.1.1
Low expression level

Low protein expression is often one of the most severe problems, and it may take a significant effort to find improvements. Here a number of hints is given which can help in many cases to advance the process of optimization (Table 9.1).

General. The level of expression of recombinant proteins usually decreases with the number of cell cycles. A main reason for this effect is that expression of non-native proteins, and also overexpression of native proteins, usually causes a negative selection pressure, i.e., cells with lower expression levels tend to outgrow the others. Thus, one should always avoid an unnecessarily large number of cell cycles and strong selection pressures. It may be safer to use freshly transformed cells for the cell culture, but in many cases not-too-old glycerol stocks, made after a fresh transformation, are sufficient. Often, colonies of freshly transformed cells may directly be transferred from the media plate into a flask with several liters of pre-warmed medium. By keeping the cells at the optimal temperature of growth without interruption during the period from transformation to harvest, one does not only speed up the cell growth, but may occasionally improve the expression level as well. The selection pressure towards lower expression is significantly reduced in inducible expression systems with a low basal level of expression. In these systems expression is usually induced only in the medium or at later stages of the exponential growth phase of the cell culture, i.e., after a large number of cells has grown up under conditions of very little negative selection pressure.

Protein is toxic to the host cells. Use an inducible expression system with a low basal level of expression, and induce expression in a late stage of the growth phase of the host cell culture. If this does not help, for example, because the protein is highly toxic to the host cells, change the host. For different expression systems see (Ausubel et al., 1992; Coligan et al., 1996).

Expression level differs between different cells. Colonies of freshly transformed cells sometimes differ in size or shape. Different types of colonies may display significant differences in expression level. Thus, it is a good idea to analyze the expression level of the different transformant colonies. Occasionally it is observed that the slowly growing colonies, which are visible only after a significantly longer duration of growth, display the highest expression level. Occasionally, this observation is made even for inducible expression systems.

Cells lose the plasmid. Reduce the number of cell growth cycles as described in the *General* section or insert the gene of the protein into the genomic DNA of the host cells.

Plasmid is damaged. The host cell culture may evade the pressure of high expression by selecting for a damaged plasmid with lower expression level: Subclone the gene into a new plasmid.

Protein is expressed in form of inclusion bodies. In the case of a low protein concentration in the supernatant that forms upon centrifugation of the lysed cells, check whether the pellet contains the recombinant protein. If the protein is expressed in the form of inclusion bodies, dissolve the inclusion bodies with a denaturant, e.g., urea, and slowly refold the protein by removing the denaturant, e.g., by dialysis. To minimize aggregation and misfolding, the refolding of the protein should proceed in a sufficient volume of solution. A method for increasing the yield of correctly folded protein is described in Sect. 9.3 (Fig. 9.9).

Lysozyme causes co-aggregation. Under some experimental conditions, lysozyme is prone to aggregation, especially at high temperature and during thawing after flash-freezing. When using lysozyme for the harvest, co-aggregation with the recombinant protein may cause losses. The application of ultrasonication, instead of lysozyme, for cell lysis may improve the protein preparation.

Table 9.1. Clues which might help to solve expression problems.

Problem	Possible solution
Protein is toxic to the host cells.	Use a different, inducible expression system with a low basal level of expression, and induce expression in a late stage of the cell growth phase.
Expression level differs between different cells.	After transformation, analyze the expression level of the different types of colonies, including that of the slowly growing ones.
Cells lose the plasmid.	Use only freshly transformed cells, or insert the gene into the genomic DNA of the host cells.
Plasmid is damaged.	Subclone the gene into a new plasmid.
Protein is expressed in form of inclusion bodies.	Dissolve the inclusion bodies with denaturant, then slowly refold the protein.
Lysozyme causes co-aggregation.	Use ultrasonication for cell lysis.

9.1.2
Expression errors

There are two main error sources which lead to the expression of a wrong recombinant protein (Table 9.2): 1. A wrong gene is used, for example, because of (a) wild-type contamination in preparations of mutants, (b) a polymerization error during DNA amplification, or (c) the use of impure oligonucleotides. 2. Incorrect post-translational modifications of the protein (for common types of alterations of the protein after assembly of the polypeptide chain on the ribosome see below). To rule out both error sources it is obviously not sufficient to sequence the gene.

Wild-type contamination. The frequency of this problem depends on the method of mutagenesis and on the care taken to ensure clean preparations of DNA. For example, in the PCR-mediated mutagenesis presented in Sect. 8.3.1.2, this problem is generally absent if one ensures (a) an excellent separation of digested and undigested plasmid in the preparation of the cloning vector, for example, by using a very clean gel chromatography (applies also for cassette mutagenesis), and (b) a clean separation of PCR product and template DNA.

Point mutations. Methods of mutagenesis which involve the use of polymerases may introduce additional point mutations. Error rates per nucleotide per extension, P_P, of most of the thermostable DNA polymerases are between 10^{-6} and 10^{-4} under optimal reaction conditions, corresponding to a less than 3% chance of one or several errors occurring in one extension reaction for a gene of 300 base pairs in length. In the PCR, the error grows with the number of extension reactions, z. The chance, P, that a randomly selected DNA molecule of the reaction product is error-free is

$$P \approx (1 - P_{\mathrm{p}})^{n(z-0.5)} , \qquad (9.1)$$

where n is the number of base pairs of the amplified gene. Thus, in the case of a large number of cycles of the PCR, and for the amplification of long DNA fragments, one should use polymerases with low error rates. Polymerases with low error rates typically have lower incorporation rates and a proofreading ability where the polymerase can correct misincorporated nucleotides in the growing strand of DNA being synthesized.

Another common source of undesired point mutations is the use of impure oligonucleotides for synthetic cassettes or as primers in the PCR. The error rates of synthetic oligonucleotides rapidly increase with the length of the oligonucleotides. There are various techniques for cleaning primers by using chromatography (see, e.g., Ausubel et al., 1992). A simple time-saving method is to use for hybridization an annealing temperature just low enough to enable binding of primers with the correct sequence, but too high to permit a significant binding of primers with additional point mutations.

Obviously, the adjustment of the annealing temperature is very important: Rapid cooling to a too low temperature usually increases the fraction of unspecific binding of contaminations; at too high annealing temperatures, the hybridization of the primers is less efficient, and undesired side-reactions may be more favored.

In PCR and cassette mutagenesis, and many other DNA reactions, it is of great importance to use oligonucleotides of sufficient length. Oligonucleotides that are too short will bind poorly, and thus unspecific binding and binding of contaminations will be more favored.

Point mutations are usually detected by DNA sequencing (Ausubel et al., 1992). High-resolution mass-spectrometry may provide a high probability for the absence of undesired mutations in a protein: Among the 380 possibilities for the mutation of one of the 20 natural amino acid residues to another, there are only four cases which introduce a mass change of less than 0.9 dalton under measurement conditions in which the sidechains are neutralized (see Sect. 2.1): Leu→Ile, Ile→Leu, Gln→Lys, and Lys→Gln.

Post-translational modifications. Problems connected with post-translational modifications are relatively common and obviously cannot be detected by sequencing the gene, but rather by mass-spectrometry, and occasionally also by chromatography. Examples for post-translational modifications are:

1. Partial digestion of the protein by proteases.
2. Attachments of additional amino acid residues to the polypeptide chain, for example, of a methionine to the N-terminus.
3. Covalent attachment of prosthetic groups. An example is horse heart cyto-chrome c in which the heme group is attached to the protein via two thioether bonds involving two cysteine residues.
4. Glycosylation. Carbohydrates are attached to the protein, usually via hydroxyl groups of threonine or serine or via the amine group of the asparagine or lysine sidechain.
5. N-terminal acetylation. Acetylation of the α-amino terminus usually blocks the degradation of the protein by the energy-dependent breakdown pathway in eucaryotes.
6. Phosphorylation. The attachment of a phosphate group to the hydroxyl group of threonine, serine or tyrosine, to the ε-amino group of lysine, or to the ring nitrogen of histidine usually serves as a message signal.
7. Adenylation of the hydroxyl group of a tyrosine.
8. Methylation of the lysine amino group.
9. Hydroxylation of lysine or proline.
10. Disulfide bond formation.

Table 9.2. Methods for the detection of expression errors.

Problem connected with:	Method of detection
Wild-type contamination	Sequencing, or restriction digest if the desired mutation introduces or removes an additional restriction site
Point mutation	Sequencing, and in most cases, but not always, detectable by mass-spectrometry[a]
Post-translational modifications	Mass-spectrometry for almost all cases[a]

[a] The mass assignment accuracy of electrospray ionization mass spectrometers is usually in the range of 0.001−0.05% for a protein of up to 100 kDa molecular weight. Note, that in exceptional cases, two errors that occur simultaneously have no detectable effect on the mass.

9.2
Aggregation

9.2.1
Detection

Aggregation of proteins is not always connected with flocculation and precipitation that is observable with the naked eye. To rule out the occurrence of aggregation in a protein sample, more sensitive methods must be used. In kinetic folding experiments transient aggregation might occur, i.e., aggregates might form only during the folding reaction and vanish thereafter. Transient aggregates may easily be confused with folding intermediates.

Absorption spectroscopy. Aggregation causes increased light scattering (Fig. 9.1). This scattering gives rise to an extinction at wavelengths where the protein usually does not absorb. For example, many proteins do not have chromophores for wavelengths around 400 nm. The extinction caused by light scattering rapidly increases with decreasing wavelength which causes a characteristic shape of the spectrum (Fig. 9.1). For proteins which absorb around 400 nm, often a different wavelength suitable for the observation of scattering can be found.

Dynamic light scattering. The scattering of small aggregates, for example dimers or trimers, is often too small to be easily detectable by absorption spectroscopy. Dynamic light scattering (Fig. 9.2) is generally more sensitive to changes of the association state of macromolecules. In this method the rotation correlation time of the molecular tumbling is used to determine the molecular weight of the particles. For proteins of >10 kDa, typically quantities of 0.01–1 mg protein dissolved in 10–100 μL buffer are needed.

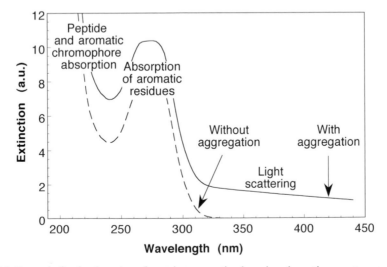

Fig. 9.1. Example for the detection of protein aggregation by using absorption spectroscopy.

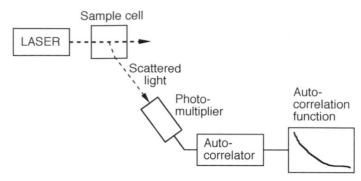

Fig. 9.2. Principle of dynamic light scattering. Light is scattered from a solution of macromolecules. The light intensity is measured as function of time and the autocorrelation function of the intensity fluctuations is calculated. Fluctuations of the scattered light intensity are related to the molecular weight of the scattering molecules. The angular dependence of scattering contains information about the shape of the macromolecules. Dynamic light scattering is an exquisitely sensitive method for the detection of aggregation.

Ultracentrifugation. Upon ultracentrifugation, the protein molecules move towards the bottom of the centrifugation flask (Fig. 9.3). Because the Brownian motion acts against this movement, the protein concentration close to the bottom will approach an equilibrium. From the concentration profile in equilibrium, measured by absorption spectroscopy, the concentrations of monomers, dimers, and multimers are determined.

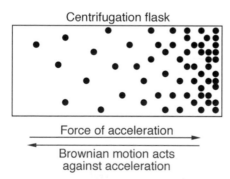

Fig. 9.3. Detection of aggregation of proteins by using ultracentrifugation. The sedimentation equilibrium is monitored at different positions of the centrifugation flask by using absorption spectroscopy. The light beam (not shown) is perpendicular to the force of acceleration. Larger particles tend to move closer to the bottom of the flask.

Kinetic analysis. In order to prove that kinetic folding experiments are not affected by aggregation, it is not sufficient to show that the observed rate constant is concentration-independent over a few orders of magnitude of concentration. For example, the observed relaxation may be caused by, or affected by, the rapid dissociation of slowly forming aggregates, which is generally a concentration-independent process. To provide further indication for the absence of these complications, it is useful to measure also the relative amplitude of the kinetic event as function of protein concentration (see Sect. 8.3.3.4).

9.2.2
Avoidance of aggregation

In general, the magnitude of aggregation and structure of aggregates strongly depends on a number of chemical and physical factors:

Protein concentration. Obviously, the magnitude of protein association events depends on the protein concentration, and thus, may usually be diminished by lowering the protein concentration. For a simple monomer–dimer equilibrium:

$$2F \; \underset{k_{-1}}{\overset{k_1}{\rightleftharpoons}} \; F_2 \, , \tag{9.2}$$

the equilibrium concentrations of monomer, $[F]_{eq}$, and dimer, $[F_2]_{eq}$, are given by Eq. 9.3 (see Fig. 9.4 and Sect. 4.5):

$$[F]_{eq} = \sqrt{\frac{k_{-1}^2}{16k_1^2} + \frac{[F_{tot}]k_{-1}}{2k_1}} - \frac{k_{-1}}{4k_1} \tag{9.3}$$

$$[F_2]_{eq} = 0.5\,([F_{tot}] - [F]_{eq})$$

where $[F_{tot}]$ is the total protein concentration, expressed in units of monomers. Practically, the concentration of dimers becomes significant for about (Fig. 9.4):

$$[F_{tot}]k_1/k_{-1} > 0.1 \tag{9.4}$$

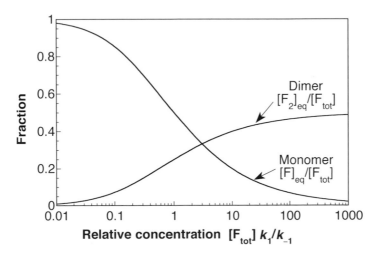

Fig. 9.4. Concentration of monomers, $[F]_{eq}$, and dimers, $[F_2]_{eq}$, as function of the total protein concentration, $[F_{tot}]$, in a monomer–dimer equilibrium, $2F \rightleftharpoons F_2$. k_1 and k_{-1} are the forward (on-) and backward (off-) rate constant, respectively (see Eq. 9.2).

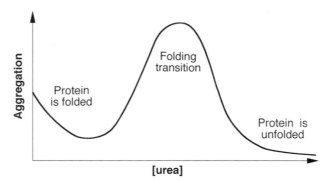

Fig. 9.5. Example for the protein aggregation as function of urea concentration. Moderate and very high concentrations of urea may often dissolve aggregates. At urea concentrations where the protein starts to unfold, sometimes increased aggregation is observed.

Denaturant concentration. Aggregation may often be reduced by adding small amounts of denaturants, for example urea. When further increasing the concentration of denaturant, most proteins will partially unfold which occasionally leads to increased aggregation (Fig. 9.5). At very high concentrations of denaturant where the protein is mainly unfolded, usually the aggregation decreases again.

Co-solvents. Co-solvents, such as ethanol, propanol, trifluoroethanol, often increase aggregation levels, mainly because they modify the hydration shell of the protein molecule and tend to enhance the intermolecular hydrophobic interaction, which is a main driving force of many aggregation processes.

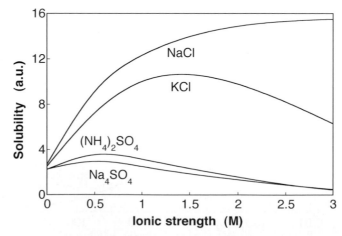

Fig. 9.6. Solubility of carbon monoxide horse hemoglobin in different salt solutions as function of the ionic strength (Cohn and Edsall, 1942). Similar to observations for many other proteins, addition of moderate concentrations of these salts causes increased solubility of hemoglobin. However, some salts decrease the protein solubility at very high concentrations of salt (salting out effect).

Surfactants. Addition of a variety of surfactants, for example, simple alkanols and micelle-forming surfactants, can often reduce aggregation and substantially increase renaturation yields of unfolded proteins (see Wetlaufer and Xie, 1995).

Salt-concentration. Aggregation is not only affected by hydrophobic interactions but also by electrostatic forces. For example, occasionally aggregation may be reduced by engineering charge mutations which cause a stronger charge repulsion between the protein molecules at low salt concentration. In those cases for which the charge repulsion between the protein molecules is important for the suppression of aggregation, the use of denaturants with high ionic strength (e.g., guanidine hydrochloride) in experiments for the study of folding is problematical. On the other hand, the solubility of many proteins decreases at very low ionic strength (Fig. 9.6).

pH. At the isoelectric point (pH = pI) the net charge of the protein is zero. In the vicinity of this point, often a particularly high tendency for aggregation is found (Fig. 9.7).

Temperature. Below 100°C, temperature elevation leads to a strengthening of hydrophobic interactions (Sect. 3.4) which often causes increased aggregation.

Chaperonins. Chaperonins, for example GroEL and GroES (Goloubinoff et al., 1989a, b; Martin et al., 1993, 1994; Hunt et al., 1996), are proteins and protein complexes which facilitate folding and can often reduce misfolding and aggregation *in vivo* and *in vitro*.

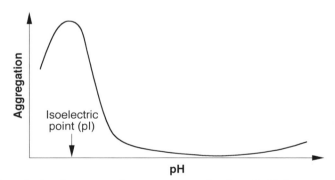

Fig. 9.7. Example for the course of aggregation as function of pH for a protein which is significantly charged at pH > pI. Vastly different curves are observed for different proteins.

9.3
Misfolding

Usually, the energy landscape of a protein under folding conditions displays a global minimum, which is occupied by the folded conformation, and local minima which may trap the protein into misfolded conformations if the folding reaction proceeds under unfavorable conditions (Figs. 1.3 and 9.8). Only in very rare cases is the native conformation not at the global energy minimum (Sohl et al., 1998).

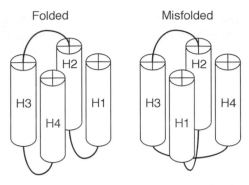

Fig. 9.8. Example for misfolding. Misfolding is the occurrence of an incorrectly folded protein conformation which has a stability that is usually lower than the stability of the folded (native) conformation, but whose rate of transition to the correctly folded conformation is very slow. The folded and misfolded conformations in this example do not necessarily differ significantly in stability or surface burial.

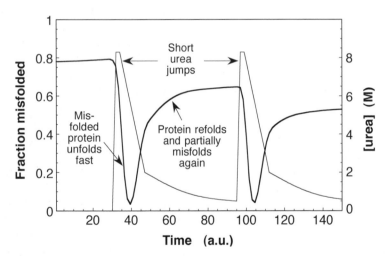

Fig. 9.9. Reduction of misfolding and aggregation by using unfolding-refolding cycles. Some preparations of overexpressed recombinant proteins yield low fractions of correctly folded protein. In many cases the property of a misfolded protein usually to unfold faster than the correctly folded protein may be used for improved reconstitution. Incubating the protein for a short period of time under unfolding conditions causes unfolding of a larger percentage of the misfolded protein than of the correctly folded protein. In the subsequent refolding reaction, a fraction of the unfolded (previously mainly misfolded) protein folds into the native conformation. Correctly folded protein accumulates in these unfolding–refolding cycles.

A simple method to decrease the fraction of misfolded protein in preparations of expensive proteins is shown in Fig. 9.9. This method makes use of the property of most misfolded proteins to unfold faster than the correctly folded protein:

1. The protein is incubated for a certain period of time under conditions that favor unfolding. This period is long enough to enable unfolding of a large fraction of misfolded and aggregated protein, but too short to enable a significant degree of unfolding of correctly folded protein.
2. To refold the protein, the denaturant concentration is lowered, for example, by mixing with a buffer.
3. Because again only a fraction refolds correctly, steps 1 and 2 are repeated several times. Prior to return to step 1, aggregates are collected, e.g., by centrifugation, and the soluble protein is concentrated by ultra-filtration.

9.4
Unstable curve fit

Many of the methods presented rely on fitting equations to experimental data. However, one may easily generate huge errors if equations with too many free parameters are fitted to curves of simple shape. This is shown for the example of the determination of the Gibbs free energy change upon folding (Fig. 9.10): Table 9.3 displays the results for curve fits with two different equations: Fit 1 is made according to Eq. 8.13 (Sect. 8.3.2) which contains slopes for the folded and unfolded state. Fit 2 is made with the same equation, but using no slope for the folded state ($\beta_F = 0$). Data set 1 contains all the points shown in Fig. 9.10. In data set 2 the point at zero concentration of urea, indicated by a square in Fig. 9.10, is left out. The two fits with slope for the folded state for the complete and incomplete set of data, respectively, are shown in Fig. 9.10. Even though the two fits are almost superimposable, the results are quite different. The use of a slope for the folded state makes the fit quite unstable: The removal of a single data point changes the result by 0.8 kJ mol^{-1} (0.2 kcal mol^{-1}). Better stability is obtained when using fewer free parameters. In this case the slope for the folded state is very small and might be neglected or be fixed (Fit 2 in Table 9.3). This changes the result for the Gibbs free energy, but for the Φ-value analysis (Sect. 8.3) only differences in Gibbs free energies between mutants and wild-type protein are needed. When using the same type of equation for all curve fits, usually the systematic errors for the energies will partially cancel each other out.

Table 9.3. Gibbs free energy change upon folding, ΔG_{F-U}, obtained for two different fits and two different data sets (see text and Fig. 9.10). (1 kJ mol^{-1} = 0.24 kcal mol^{-1})

	Fit 1 (with slope for the folded state, shown in Fig. 9.10)	Fit 2 (with no slope for the folded state)
Data set 1 (all data)	−14.7 kJ mol^{-1}	−13.9 kJ mol^{-1}
Data set 2 (without point at zero urea concentration)	−15.5 kJ mol^{-1}	−13.6 kJ mol^{-1}

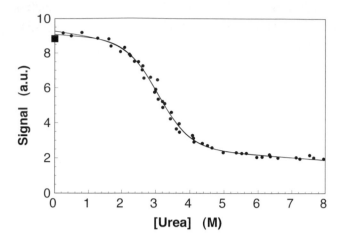

Fig. 9.10. Determination of the Gibbs free energy change upon folding from an equilibrium denaturant titration curve of a protein by using an equation (Eq. 8.13 in Sect. 8.3.2) that contains a large number of free parameters. The two, almost superimposable, fits for the complete and incomplete data set, respectively, are shown. In the incomplete data set one point is removed at zero concentration of denaturant. Quite different Gibbs free energy changes upon folding, ΔG_{F-U}, are obtained for the two different data sets when using too many free parameters for the curve fit. A more stable fit is obtained in this example when using no slope or a fixed slope for the signal of the folded state (see Table 9.3). When comparing energy differences between mutants and wild-type, the introduced systematic errors partially cancel each other out.

10 The folding pathway of a protein (barstar) at the resolution of individual residues from microseconds to seconds

10.1 Introduction

This chapter reports the first resolution of a folding pathway of a protein (barstar, the 10 kDa polypeptide inhibitor of the ribonuclease barnase, see Fig. 10.1) from a well-characterized unfolded state (Sect. 10.4) at the level of individual amino acid residues on a microsecond to second time scale. The presentation, which is largely based on several articles by the author and coworkers (Nölting et al., 1995, 1997a, b; Nölting, 1996, 1998a, b, 1999; Nölting and Andert, 2000), is intended to illustrate the method of Φ-value analysis (see Sect. 8.3) in detail and in a more applied way than in the previous chapters.

A number of studies (Nölting et al., 1995, 1997a, b; Nölting, 1998a) has shown that a folding intermediate with marginal stability, 2.5 kJ mol^{-1}, is formed within a few hundred µs after rapidly raising the temperature of the (partially) cold-unfolded solution of barstar from 2 to 10°C (Sect. 10.5). Φ-value analysis (Sects. 10.6–10.8) shows that the first and fourth helix become substantially consolidated as the intermediate is formed, stabilized by long-range interactions. A native-like structure is then formed within about 100 ms as the whole structure consolidates. The overall folding pathway fits the nucleation–condensation model in which structure is (partly) formed in a diffuse nucleus and then becomes consolidated as further structure condenses around the nucleus or as several modules of structure dock (Nölting et al., 1997a). The structure of the diffuse nucleus and its growth in three stages, 500 µs, 1 ms, and 100 ms after initiation of the folding reaction, is mapped out by correlating Φ-values for a set of single mutants with inter-residue contact charts (Sects. 10.7, 10.8; Nölting, 1998a, 1999). The folding nucleus detected in barstar initially comprises mainly secondary and tertiary interactions of strand$_1$, loop$_1$, and helix$_1$ (Nölting et al., 1997a; Nölting, 1998a). The discovered highly anisotropic folding behavior may explain the high speed of the folding reaction, compared with the speed of a random sampling. In agreement with the nucleation–condensation model of folding (Sect. 10.11), the non-uniform structure consolidation is most pronounced in the early stages of folding. The late folding events of barstar are characterized by a propagation of structure consolidation from the N- and C-termini towards amino acid residues located close to the center of the polypeptide chain (Nölting, 1998a).

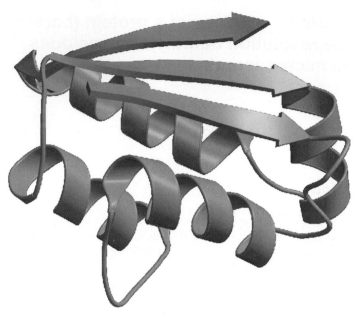

Fig. 10.1. Folded structure of barstar, drawn with the program MOLSCRIPT (Kraulis, 1991) and Raster3D (Bacon and Anderson, 1988; Merritt and Murphy, 1994).

10.2
Materials and methods

Protein expression and purification. The "pseudo-wild-type" barstar used in the studies which are presented in Sects. 10.3–10.11 is C40A/C82A/P27A barstar containing no cysteines and only one proline residue that is at position 48 (Nölting et al., 1995, 1997a, b; Nölting, 1998a). Mutants were engineered by cassette mutagenesis or by PCR mediated mutagenesis (see Sect. 8.3.1 and Ausubel et al., 1992). Screening of mutant plasmids (pML2bs), expressed in *Escherichia coli* BL21(DE3)(pLysE), was simplified by introducing a silent mutation for a restriction site. For expression, freshly transformed cell colonies were transferred directly into a 2-L flask with 800 mL 2-TY medium with 200 µM ampicillin and incubated at 30°C for 30 h with addition of 4 mM isopropyl-1-thio-β-galactoside (IPTG) at 10 h and 20 h each. Intense pulsed ultrasonication of the ice-cold cell-suspension was used instead of lysozyme for cell lysis, resulting in >95% purity of barstar mutants in the extracted inclusion bodies before ion-exchange chromatography. Electrospray mass spectrometry confirmed the size of the mutants within ±0.8 dalton. Mass spectrometry and N-terminal sequencing showed that the N-terminal methionine was not cleaved in these highly expressed barstar mutants, including pseudo-wild-type. The expression level was typically 100 mg L^{-1}, and the protein was concentrated to 1–2 mM in water without any precipitation occurring (Nölting et al., 1995, 1997a, b; Nölting, 1998a).

Circular dichroism (CD) spectra. CD was measured using a JASCO (Easton, MD) model J-720 spectropolarimeter that was interfaced with a computer-controlled Neslab (Newington, NH) RTE-111 waterbath. CD calibration was performed using (1S)-(+)-10-camphor-sulfonic acid (Aldrich, Milwaukee, WI) with a molar ellipticity of 2.36 L mol^{-1} cm^{-1} at 290.5 nm and a molar extinction coefficient of 34.5 L mol^{-1} cm^{-1} at 285 nm. CD spectra were obtained with a spectral resolution of 2 nm, an optical pathlength of 1 cm and a protein concentration of about 5 μM. For each spectrum, 16 scans of about 1 minute duration were accumulated and averaged. Barstar concentrations were determined using an extinction coefficient at 280 nm of 22,690 L mol^{-1} cm^{-1} (Lubienski et al., 1994; Nölting et al., 1995).

CD temperature scans. CD temperature scans were done with a heating rate of 50°C h^{-1}. The absolute error of temperature is ±1.5°C and the relative error of temperature between curves in one diagram is less than ±0.3°C. The absolute error of the CD signal, resulting from the error in the determination of the protein concentration and errors in the CD calibration, is less than ±6% at 25°C. The reversibility of thermal unfolding in 3 M urea was only about 85%, but the reversibility of the temperature-induced changes in the unfolded state was better than 99.5%, as judged by heating a 5-μM barstar sample in 7 M urea at 20°C h^{-1} from 1°C to 95°C, then cooling down to 1°C and heating again to 95°C at 60°C h^{-1}. CD measurements shown in each diagram were done with the same batch of protein and sample cell. The relative error of $\Delta\varepsilon_{R,230}$ for the scans on each side of Fig. 10.4 is ±0.01 L mol^{-1} cm^{-1} ±0.01×$\Delta\varepsilon_{R,230}$ as judged by the reproducibility. No differences for different temperature scan rates above 20°C were noticed. Below 20°C, the systematic error of $\Delta\varepsilon_{R,230}$ due to kinetic effects is up to −0.02 L mol^{-1} cm^{-1}. Temperature scans were started at a temperature 5°C lower than those shown in Fig. 10.4 (Nölting et al., 1995, 1997a, b).

Stopped-flow circular dichroism studies. Kinetic stopped-flow experiments used an Applied Photophysics BioSequential DX-17 MV stopped-flow spectrometer complemented with a CD.1 circular dichroism accessory (Leatherhead, UK) with a dead time of 7 ms. Refolding was initiated by 11-fold dilution of urea solutions with 260 μM protein. Measurements of the changes in CD at 222 nm were performed at 15°C in 10 mM sodium phosphate buffer at pH 8.0. Only a single slow rate constant of 3–10 s^{-1}, depending on urea concentration, was observed. Its amplitude was obtained as a function of the concentration of urea by fitting the refolding curves to a single-exponential function. A small correction was made for the amplitude lost in the dead time. The denaturant dependence of the amplitude (Fig. 10.12) mainly reflects the stability curves of the secondary structure of the intermediate and native-like folded state with *trans*-conformation of the peptidyl-prolyl (48) bond, respectively (Nölting et al., 1997a).

Fluorescence spectra. For equilibrium fluorescence experiments with the fluorescence excitation at 280 nm, a sample cell of 0.4 cm × 1 cm and a protein concentration of 4 μM were used. At 2°C, the sample was equilibrated for 2 h prior to the measurement (Nölting et al., 1995).

Equilibrium unfolding. Free energies of unfolding for pseudo-wild-type and mutants in 0 M urea, ΔG_{F-U}, were determined as described in Sect. 8.3.2 by using CD. The CD signal at 222 nm for protein samples in 50 mM Tris-HCl buffer at pH 8 with 100 mM KCl at 10°C was detected as a function of urea concentration using a protein concentration of 10–20 μM and an optical pathlength of 0.1 cm, and the spectral resolution was adjusted to 2 nm (Nölting et al., 1995, 1997a, b; Nölting, 1998a).

T-jump measurements. To avoid artifacts, only small-amplitude T-jumps from 2°C to 10°C and low protein concentrations were used with a T-jump fluorimeter from DIA-LOG (Düsseldorf, Germany), equipped with a 50-nF capacitor, a 200-W mercury–xenon lamp, and supplemented with a NICOLET (Madison, WI) model Pro 90 storage oscilloscope (see Fig. 5.6). The conditions were, unless stated otherwise: Fluorescence signals at ± 90° angles were detected with two photomultipliers, summed and filtered with a 5-μs response time for the fast transition and a 1-ms response time for the slow transition. 2000 data points at 12-bit resolution were recorded for each trace. Fluorescence excitation was at 280 nm, and a cut-off filter at 295 nm was used for emission. Folding proceeds under isothermal conditions since after the T-jump, the temperature of the interior of the protein molecule equilibrates with the bulk solvent in the nanosecond time scale (Nölting, 1995, 1998b). The conditions were typically 2–10 μM protein, 50 mM Tris-HCl buffer at pH 8 with 100 mM KCl, urea concentration as stated. At 2°C, about 1% of the pseudo-wild-type barstar molecules are in the unfolded state. Changing the concentration of pseudo-wild-type from 5 to 30 μM at 0 M urea does not change the rate constants and relative amplitudes of the fast folding transition by more than 5%, but deviations could be seen at protein concentrations in the mM range (Nölting et al., 1997a, b; Nölting, 1998a).

NMR studies on peptide fragments. Peptides of barstar comprising residues 11–29, 28–44, 33–44 and 14–43 were synthesized with a Synergy Personal Peptide Synthesizer, using F-moc protection. All were purified by reverse HPLC chromatography. NMR spectra of peptides at 5°C in acetate buffer at pH 5.3 were acquired using standard pulse sequences (Wüthrich, 1986) with an AMX-500 Bruker spectrometer. Peptide concentrations were in the range of 1 to 1.5 mM (Nölting et al., 1997a).

10.3
Structure of native barstar

Barstar is an 89-amino acid residue protein that has evolved to be the specific intracellular inhibitor of the ribonuclease barnase. Both are expressed from *Bacillus amyloliquefaciens* (Hartley, 1988; Schreiber and Fersht, 1993a; Nölting et al., 1995, 1997a). Barstar (Fig. 10.1) has four α-helices and three strands of parallel β-sheet; helix$_1$ from Ser14 to Ala25, helix$_2$ from Asn33 to Gly43, helix$_3$ from Gln55 to Thr63, helix$_4$ from Glu68 to Gly81, strand$_1$ from Lys1 to Asn6, strand$_2$ from Leu49 to Arg54, and strand$_3$ from Asp83 to Ser89. Residues 26–44 comprise a loop as well as helix$_2$ of barstar, forming the binding site for barnase. The inter-residue contact map (Fig. 10.2) shows that the divisions between possible subdomains, such as residues 1–50 and 51–89, are very weak and so barstar is mainly a single domain protein (Nölting, 1998a).

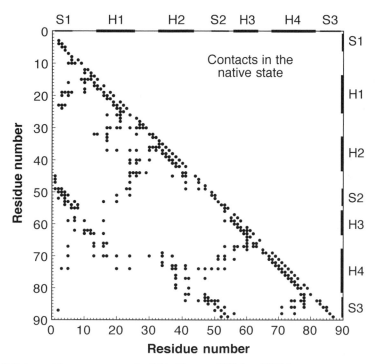

Fig. 10.2. Inter-residue contact map for folded barstar (Nölting, 1998a).

10.4
Residual structure in the cold-unfolded state

To assess the structural changes along the pathway of folding (Nölting et al., 1997a; Nölting, 1998a), it is important to know the spectroscopic properties, thermodynamics and structures of the unfolded states. In particular, it is crucial to know the differences in the residual structure between the different unfolded states. Important questions regarding unfolded states of protein remain to be answered: How can the structural distribution of unfolded protein be resolved? What is the energy landscape of unfolded protein? What is the nature of the transitions between different unfolded states? Do the transitions involve coopera-tive melting of structure or gradual changes? How might structures be funneled at the beginning of the folding reaction before reaching the first detectable transition state (Nölting et al., 1997b)?

The cold-unfolded state of barstar has been extensively analyzed by nuclear magnetic resonance (NMR) and circular dichroism (CD) spectroscopy (Wong et al., 1996; Nölting et al., 1997b). From NMR, three regions have been shown to contain residual structures which are in the α-helical region of (φ,ψ) space. Two of these are native-like since they are also helical in the native protein: Ser12 to Lys21 of helix$_1$; Tyr29 to Glu46 (which contains helix$_2$). The third, Leu51 to Phe56, is part of strand$_2$ plus the loop connecting it to helix$_3$ and so is non-native. The C-terminal region, Asn65 to Ser89 (which contains helix$_4$ and strand$_3$), is indistinguishable from a random coil (Nölting et al., 1997a).

Consistent with the NMR data, the far-UV CD of unfolded pseudo-wild-type (C40A/C82A/P27A) barstar displays a surprising sensitivity to the point mutations Q18G and A25G, located in helix$_1$ in the native structure (Fig. 10.3; Nölting et al., 1997b). In contrast, the chemically similar mutants Q72G and A77G, located in helix$_4$, and Q58G, located in helix$_3$, display little, if any, difference of the far-UV CD signal relative to pseudo-wild-type. Similarly, the CD at 230 nm of S14A in helix$_1$ is more negative than that of S59A in helix$_3$ (Nölting et al., 1997b).

The dependence of the CD signal of the unfolded state on the concentration of urea (Nölting et al., 1997b) and on the temperature (Fig. 10.4; Nölting et al., 1997b) indicates that the residual structure in unfolded barstar is of very low stability and does not involve significant surface burial. For example, no cooperative transitions in the unfolded state in high concentrations of urea are observed in CD temperature scans (Fig. 10.4).

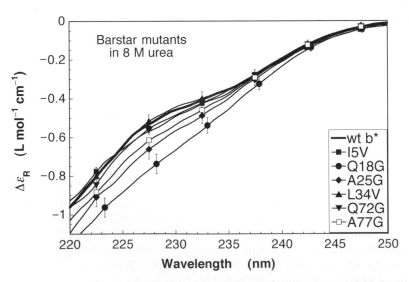

Fig. 10.3. Spectra of the mean residual ellipticity, $\Delta\varepsilon_R$, of the urea-unfolded states of C40A/C82A/P27A barstar (wt b*) and its mutants: I5V, Q18G, A25G, L34V, Q72G and A77G (Nölting et al., 1997b). Different symbols are used for better visibility and do not reflect the number of points measured.

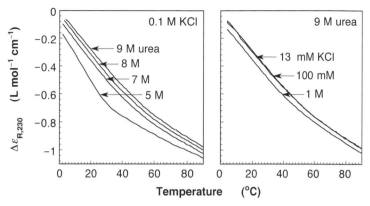

Fig. 10.4. CD temperature scans of C40A/C82A/P27A barstar at different concentrations of urea and KCl, as indicated (Nölting et al., 1997b).

10.5
Gross features of the folding pathway of barstar
10.5.1
Equilibrium studies

The C40A/C82A/P27A (pseudo-wild-type) barstar displays cold-unfolding at moderately low temperatures ($\approx 0°C$; see Fig. 10.5). There is a significant loss of far-UV signal upon cold- and heat-unfolding, indicative of the melting of the secondary structure (Nölting et al., 1995).

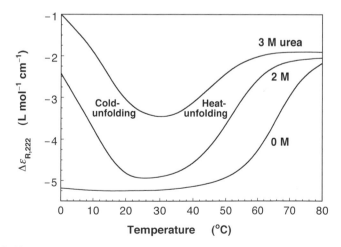

Fig. 10.5. Cold- and heat-unfolding of C40A/C82A/P27A barstar at different urea concentrations, as indicated (Nölting et al., 1995).

The equilibrium unfolding curves on addition of urea are identical when monitored by far-UV CD at 222 nm, which follows the secondary structure, and fluorescence emission at 330 nm, which follows tertiary structure interactions (Fig. 10.6). The m-value ($m = \partial(\Delta G_{F-U}[\text{urea}])/\partial[\text{urea}]$, where $\Delta G_{F-U}[\text{urea}]$ is the Gibbs free energy change upon folding at various concentrations of urea, [urea]) is as high as that for wild-type barstar, at 5.0–5.4 kJ L mol^{-2} (1.2–1.3 kcal L mol^{-2}) and the population of any intermediates at equilibrium is small. Both heat- and cold-induced unfolding are cooperative (Nölting et al., 1995, 1997a).

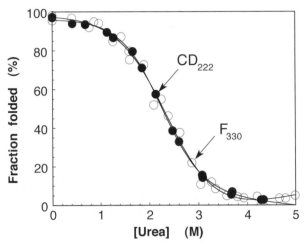

Fig. 10.6. Urea-induced unfolding of C40A/C82A/P27A barstar, monitored by fluorescence emission at 330 nm (F_{330}) with excitation at 280 nm and CD at 222 nm (CD_{222}), as indicated (Nölting et al., 1995).

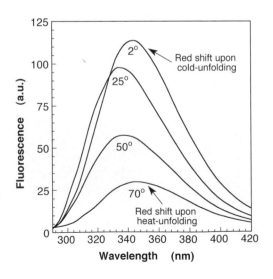

Fig. 10.7. Cold- and heat- unfolding of pseudo-wild-type bar-star in 2 M urea, monitored by fluorescence (Nölting et al., 1995).

The fluorescence spectrum (Fig. 10.7) displays a significant red-shift upon cold and heat-unfolding, indicative of the solvent exposure of aromatic side-chains upon unfolding (see also next section; Nölting et al., 1995).

10.5.2
Kinetic studies

There is a *cis*-peptidyl-prolyl bond to the proline at position 48, which complicates the folding pathway since the major conformation present in the unfolded state, U, is the *trans*. The major pathway for the folding of barstar is the sequence (Nölting et al., 1995, 1997a; Nölting, 1998a):

$$U_{trans} \underset{k_{-1}}{\overset{k_1}{\rightleftarrows}} I_{trans} \underset{k_{-2}}{\overset{k_2}{\rightleftarrows}} F_{trans} \underset{k_{-3}}{\overset{k_3}{\rightleftarrows}} F_{cis} \qquad (10.1)$$

where I_{trans} is the early intermediate and F_{trans} is a native-like state which binds barnase and has similar fluorescence properties to the fully folded state, F_{cis}. The sequence $I_{trans} \rightleftarrows F_{trans} \rightleftarrows F_{cis}$ has been detected by stopped-flow studies (Schreiber and Fersht, 1993b; Agashe and Udgaonkar, 1995; Nölting et al., 1997a; Nath and Udgaonkar, 1997b). It was found from temperature-jump studies that I_{trans} is a state that has a compactness in-between that of the unfolded and folded states: On the one hand it is clearly more collapsed than the unfolded state, on the other hand the solvent exposure of aromatic sidechains is significantly larger than in the folded state (Nölting et al., 1995, 1997a).

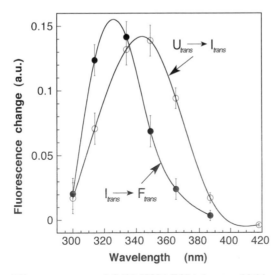

Fig. 10.8. Kinetic difference spectra of C40A/C82A/P27A barstar (Nölting et al., 1995). The amplitudes of the relaxations are measured as function of the wavelength of fluorescence emission for the transitions $U_{trans} \rightarrow I_{trans}$ and $I_{trans} \rightarrow F_{trans}$, as indicated.

There is evidence from the analysis of spectral changes for the burial of hydro-phobic side chains in the slow transition, I→F (Nölting et al., 1995). The fluores-cence spectrum of pseudo-wild-type barstar shows a characteristic blue shift upon cold- and heat-refolding (Fig. 10.7), indicative of the decrease in solvent exposure of the tryptophan side chains on folding. The significant increase in fluorescence accompanied by a blue shift during the slow kinetic event (Fig. 10.8) indicates a burial of the side chains of tryptophan. In contrast, the shape of the difference spectrum of the fast transition, U→I, derived from monitoring the amplitudes of the transition between 300 and 400 nm, resembles that of the equilibrium spectrum of the cold-unfolded protein (Fig. 10.7), although there is also an increase in amplitude. At a wavelength of 418 nm, however, the amplitude of the kinetic difference spectrum (Fig. 10.8) is negative, suggestive of a small blue shift. These observations indicate an increase in hydrophobic burial of tryptophan side chains for the fast phase, too (Eftink and Shastry, 1997). The position of the wavelength maximum in the difference spectrum, however, suggests that the fast transition occurs between solvent-exposed states (Nölting et al., 1995).

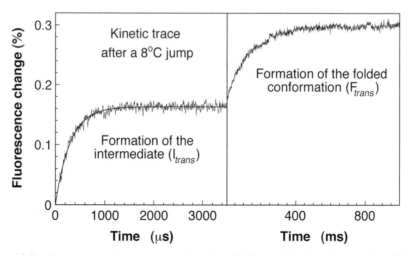

Fig. 10.9. Temperature-jump trace for the folding of 6 μM pseudo-wild-type (P27A/C40A/C82A) barstar in the absence of denaturants at 10°C (Nölting et al., 1995, 1997a). 42 traces were accumulated with time constants of 5 μs and 1 ms for the fast ($U_{trans} \rightleftharpoons I_{trans}$) and slow ($I_{trans} \rightleftharpoons F_{trans}$) transition, respectively. There is no evidence for a lag phase preceding the formation of I_{trans}. The constant for the slow transition matches with that obtained using rapid mixing.

When monitoring folding from the combined changes in fluorescence of tryptophans 38, 44, and 53, it is found that the folding intermediate I_{trans} is formed with a rate constant $k_1 = 2300$ s^{-1} and decays with $k_{-1} = 800$ s^{-1} at 10°C (Figs. 10.9–10.11), so that $\Delta G_{I_{trans}-U_{trans}} = -2.5$ kJ mol^{-1} (-0.6 kcal mol^{-1}). The

I$_{trans}$ proceeds to F$_{trans}$ with $k_2 = 11$ s^{-1}, and k_{-2} is about 2 s^{-1}, so that $\Delta G_{F_{trans}-I_{trans}} = -4.0$ kJ mol^{-1} (-1.0 kcal mol^{-1}). The overall Gibbs free energy change upon folding is -12.5 kJ mol^{-1} (-3.0 kcal mol^{-1}), so that $\Delta G_{F_{cis}-F_{trans}}$ is -6.0 kJ mol^{-1} (-1.4 kcal mol^{-1}). The conversion of F$_{trans}$→F$_{cis}$ has a half life of several minutes at 10°C (Nölting et al., 1995, 1997a).

The change in solvent exposure of each state during folding has been estimated from the sensitivity of the folding and unfolding rate constants of both transitions to the effect of the urea concentration, [urea], (Fig. 8.26) compared with that of the overall equilibrium constant for folding, K_{F-U} (Figs. 8.21, 8.22). For the overall folding reaction, the slope is $\partial \log K_{F-U}/\partial[urea] = -0.9$ L mol^{-1}. The individual folding and unfolding rate constants have been found to change with the urea concentration according to: $\partial \log k_1/\partial[urea] = -0.2$ L mol^{-1}; $\partial \log k_{-1}/\partial[urea] = 0.3$ L mol^{-1}; $\partial \log k_2/\partial[urea] = -0.3$ L mol^{-1}; and $\partial \log k_{-2}/\partial[urea] = 0.1$ L mol^{-1}. Thus, the compactness in surface area of the first transition state, #1, the intermediate, I$_{trans}$, and the second transition state, #2, is approximately 20%, 50%, and 90%, respectively, (Nölting et al., 1995, 1997a).

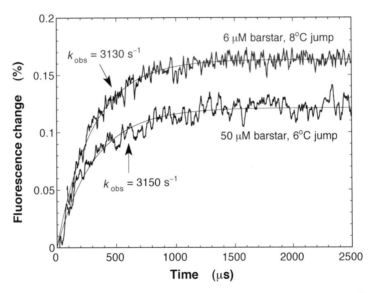

Fig. 10.10. Temperature-jump traces for the fast folding transition, U$_{trans}$⇌I$_{trans}$, of P27A/C40A/C82A barstar at 10°C under different instrument settings and concentrations (Nölting et al., 1995, 1997a). 36 traces were accumulated, each with 500 ns step width, corresponding to 5000 points per trace and a time constant of 1 μs for the measurement with 50 μM protein. Noise was reduced by digital smoothing using a moving window of 50 data points. For the measurement with 6 μM protein, 42 traces were accumulated with a time constant of 5 μs. A 5 times higher light intensity relative to the measurement with 50 μM was used. The observed rate constant, $k_{obs} = k_1 + k_{-1}$, does not deviate by more than 5% of 3100 s^{-1} over the concentration range 6 to 50 μM protein and a wide range of instrument settings and heating times.

I_{trans} is thus a fairly compact state with about 50% burial of surface area (relative to the burial of surface in the folded state, taking the unfolded state as the reference state), and with about 40% of α-helical content of the folded structure (Fig. 10.12). The formation of I_{trans} is cooperative since its rate constant is affected by many mutations all over the molecule (Sects. 10.6–10.8; Nölting et al., 1997a; 1998a).

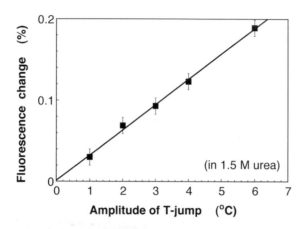

Fig. 10.11. The amplitude of the fast phase, $U_{trans} \rightleftharpoons I_{trans}$, is proportional to the amplitude of the T-jump.

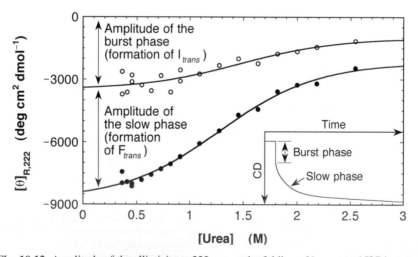

Fig. 10.12. Amplitude of the ellipticity at 222 nm on the folding of barstar at 15°C in a stopped-flow circular dichroism spectrometer (Nölting et al., 1997a). The burst of signal (open circles) in the dead time relative to the baseline corresponds to the circular dichroism signal of I_{trans}. The total amplitude of the signal on folding (closed circles) corresponds to that of F_{trans} (combined changes of the signals for the two reactions).

10.6
Φ-value analysis

The degree of formation of structure at individual positions in the intermediate and transition states, X, on the folding pathway of barstar has been estimated from the changes in their Gibbs free energy on mutation, $\Delta\Delta G_{X-U}$, relative to the change in the overall Gibbs free energy of folding, $\Delta\Delta G_{F-U}$, (Fig. 10.13; Fersht et al., 1992; Fersht, 1995a; Nölting et al., 1995, 1997a; Nölting, 1998a, 1999). For example, when the Φ-value for the state X, $\Phi_X = \Delta\Delta G_{X-U}/\Delta\Delta G_{F-U}$, is = 1, then X is destabilized by mutation by the same amount of energy as is the fully folded state, F. When $\Phi_X = 0$, X is as unaffected by mutation as is the unfolded state U. Intermediate values of Φ indicate either a mixture of states of different degrees of formation of structure or partial formation of structure (for more details see Sect. 8.3, especially Sect. 8.3.4; Fersht et al., 1992, 1994; Nölting et al., 1997a).

Figs. 10.14 and 10.15 present the Φ-values for the transition state for the formation of I$_{trans}$ (#1) and for the equilibrium constant for its formation (I), as well as those for the transition state #2 for the formation of F$_{trans}$ (Nölting et al., 1997a; Nölting, 1998a). Helix$_1$ has significant values of Φ for #1 which slightly increase in I and show that it is nearly completely formed in #2. Mutation of amino acid residues in helix$_2$ indicates that it is less well formed in #1, but these mutations are more radical than those in helix$_1$ because of the large hydrophobic

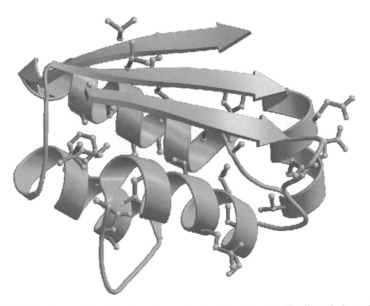

Fig. 10.13. Mutants used for the Φ-value analysis of barstar are distributed throughout the molecule; mutated sidechains are indicated as a ball-and-stick structure. Each single mutant probes the role of a part of the protein molecule in the folding reaction (Lubienski et al., 1994; Nölting et al., 1997a; Nölting, 1998a). [Figure was drawn with MOLSCRIPT (Kraulis, 1991).]

groups involved and so there are also contributions from changes in their interactions with the hydrophobic core. In particular, Leu34 which shows a high Φ in I, has strong interactions with Ser69 and Val70 in helix$_4$. Helix$_3$ is only very weakly formed in #1, I and #2. Helix$_4$ is not formed in #1, is partially consolidated in I and is significantly formed in #2 (Nölting et al., 1997a).

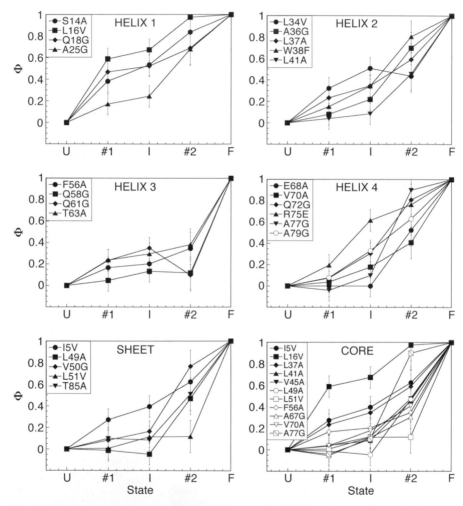

Fig. 10.14. Φ-value analysis of barstar (Nölting et al., 1995, 1997a; Nölting, 1998a). The Φ-values for the transition state for the formation of I$_{trans}$ (#1) are calculated from k_1 and those for the equilibrium constant for the formation I$_{trans}$ are calculated from the ratio k_1/k_{-1}. Φ-values for the transition state for the formation of F$_{trans}$ (#2) are calculated from the values of k_{-2} (from a folding–unfolding double-jump experiment). The error bars indicate the estimated maximum errors.

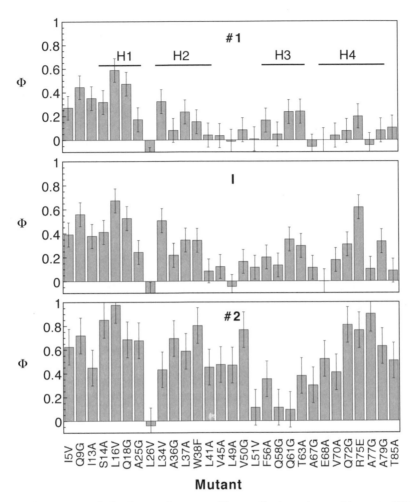

Fig. 10.15. Φ-value analysis of a set of mutants of barstar for the early transition state (#1), the intermediate (I), and late transition state (#2), as indicated (Nölting et al., 1995, 1997a; Nölting, 1998a). The error bars indicate the estimated maximum errors.

Hydrophobic amino acid residues in the core show varying degrees of structural consolidation in #1 and I (Fig. 10.14). L16V, which has a high Φ in #1 and I, mainly probes interactions in the $helix_1$–$loop_1$–$strand_1$ motif and between $helix_1$ and $helix_4$. The five probes in the strands tend to show that the β-sheet is formed primarily in I and #2 (Nölting et al., 1997a; Nölting, 1998a).

These results are further refined in the following sections. But, it is clear at this stage of refinement that the first detectable intermediate, I, is structured around $helix_1$ (being already considerably formed in #1) and $helix_4$ (being in the process of consolidation in I).

10.7
Inter-residue contact maps

In order to simplify the structural interpretation of the Φ-values it is important to use mutants which do not cause a significant disruption of the native structure outside the position of the mutation but only local changes (Matouschek et al., 1989; Fersht et al., 1992). Unfortunately it is often unavoidable that mutations probe several interactions simultaneously. In this case it is useful, for an improved interpretation of the Φ-values, to correlate the Φ-values with inter-residue contact maps (Figs. 10.16–10.18; Nölting, 1998a, 1999, 2003; Nölting and Andert, 2000).

When a mutant probes n contacts, Φ for a particular state is

$$\Phi = \sum_{i=1}^{n} (\Phi_i \times \Delta\Delta G_{F-U,i})/\Delta\Delta G_{F-U} \qquad (10.2)$$

$$\Delta\Delta G_{F-U} = \sum_{i=1}^{n} \Delta\Delta G_{F-U,i} \quad , \qquad (10.3)$$

where Φ_i and $\Delta\Delta G_{F-U,i}$ are the Φ-value and Gibbs free energy contribution upon mutation, respectively, for the contact number i, and $\Delta\Delta G_{F-U}$ is the total observed change in Gibbs free energy on mutation. For correlation with contacts, the Φ-values are equally assigned to all contacts that are predicted to be probed by mutation.

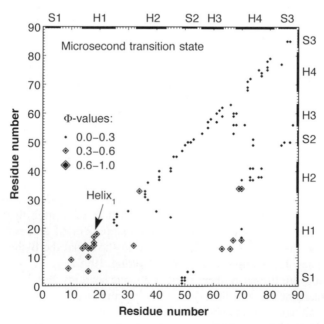

Fig. 10.16. Inter-residue contact map for the microsecond folding transition state, #1, of barstar (Nölting, 1998a).

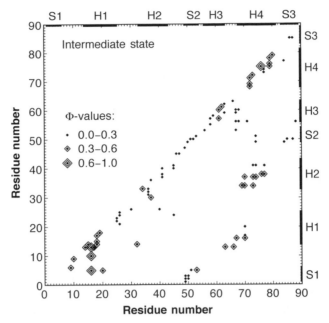

Fig. 10.17. Inter-residue contact map for the early folding intermediate, I, of barstar (Nölting, 1998a).

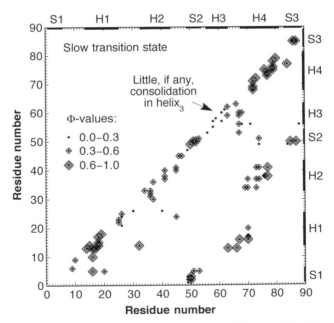

Fig. 10.18. Inter-residue contact map for the late folding transition state, #2, of barstar (Nölting, 1998a, 2003). See also Figs. 11.2 and 11.3.

Table 10.1. Predictions for the contacts which are probed upon mutation (Nölting, 1998a).

Mutant	Location	Type[a]	Interactions that are predicted to be modified by mutation (S = strand; H = helix; L = loop) [b]
I5V	strand 1	T	S2: W53; H1: L16, L20
Q9G	loop 1	S	L1: N6, I10
I13A	loop 1	T	H1: S14, L16, H17; H3: T63; L4: G66
S14A	helix 1	S	H1: D15; L2: E32
L16V	helix 1	T	S1: I5; L1: I10, I13; L4: A67; H4: V70
Q18G	helix 1	S	H1: S14, D15, H17, T19
A25G	helix 1	S	H1: K22, E23; L2: L26
L26V	loop 2	T	H1: K21, L24; L2: Y30; H2: A40
L34V	helix 2	T	H4: S69, V70, V73; H2: N33
A36G	helix 2	S	L2: G31, E32; H2: N33, L37
L37A	helix 2	T	H4: V70, V73, F74; L2: Y30
W38F	helix 2	T	H4: V73, E76, A77
L41A	helix 2	T	H2: L37, W38, A40; H4: V73, F74, A77
V45A	loop 3	T	H1: L24; H2: A40, L41; L3: E46, Y47
L49A	strand 2	T	S1: K1, K2, A3; L3: Y47; S2: V50; H4: F74; S3: I84
V50G	strand 2	T	S1: K2, A3; S2: L49, L51, E52; S3: T85, I87
L51V	strand 2	T	S1: I5; S2: W53; H4: F 74
F56A	helix 3	T	L4: A67; H4: E68, L71; S2: W53; S3: L88
Q58G	helix 3	S	H3: Q55, E57, S59
Q61G	helix 3	S	H3: E57, K60, L62
T63A	helix 3	S	L1: I13; H3: S59, L62; L4: G66
A67G	loop 4	T	S2: W53; H3: F56, S59, K60
E68A	helix 4	C	H3: K60
V70A	helix 4	T	H1: L16, H17, L20
Q72G	helix 4	S	H4: E68, S69, L71, V73
R75E	helix 4	C	H4: E76
A77G	helix 4	T	H2: W38, L41; H4: V73, F74; S3: I84
A79G	helix 4	S	H4: R75, E76, K78, E80
T85A	strand 3	T	S2: V50; S3: I86, I87

[a] S = Mainly secondary structure probing mutants (solvent-exposed sidechain); T = Mainly tertiary structure probing mutants (buried sidechain); C = Charge mutants (solvent-exposed sidechain).
[b] Predicted by using the NMR structure of wild-type barstar (Lubienski et al., 1994). Because the precise structure of the mutants is not known, this column reflects only probabilities. However, all mutants show a high activity.

In case of a Φ-value of 1 the assignment gives exact Φ-values for all individual contacts since $\Phi = 1$ can only be found if all $\Phi_i = 1$ and similarly for $\Phi = 0$ all $\Phi_i = 0$ unless there are non-native interactions with a significant Gibbs free energy contribution (Nölting, 1998a, 1999, 2003; Nölting and Andert, 2000). The occurrence of such non-native interactions is analyzed in Sect. 10.9.

When the Φ-value is in-between 0 and 1, the contact map reflects a probability of structure formation (Nölting, 1998a). When a cluster of a large number of contacts from different mutations shows consistently high Φ-values, it is reasonable to assume that the average of Φ for this cluster is non-negligible. If, for example, two contacts of about similar energy contribution are probed by a mutation, then one of the contacts can have a negligible Φ only if the average Φ is found to be 0.5 or less (Nölting, 1998a, 1999).

Further, if some mutants probe only one specific structural element E and other mutants probe the two structural elements E and F, then one can obtain a probability of structure formation in F by comparing the two sets of Φ-values. For example, the mutants in helix$_1$ L16V, Q18G, S14A, A25G with $\Phi_{\#1} = 0.59, 0.47$, 0.32, and 0.17, are predicted to probe 100%, 0%, 50%, and 33% tertiary structure contacts, respectively. No decreased Φ for a large fraction of tertiary structure contacts is obvious, suggesting a high probability for a significant consolidation of tertiary structure interactions at this stage, too (Nölting, 1998a).

On average, 3−4 intramolecular contacts are predicted to be significantly modified in the folded state upon mutation (Table 10.1). Often, several of the contacts for one mutant cluster in a relatively limited region of the inter-residue contact map (Fig. 10.16−10.18), so that often only one or two clusters of interactions are predicted to be probed by mutation. Further, to simplify the interpretation, preferentially mutants have been used which probe either only secondary or mainly tertiary structure interactions (Nölting et al., 1995, 1997a; Nölting, 1998a).

10.8
The highly resolved folding pathway of barstar

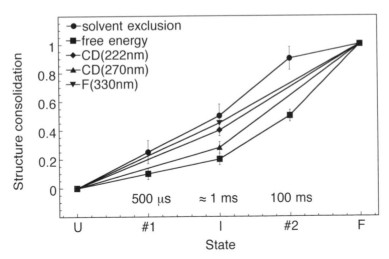

Fig. 10.19. Consolidation of structure along the folding pathway of barstar as measured by solvent exclusion (circles), Gibbs free energy (squares), CD at 222 nm (diamonds), CD at 270 nm (upward triangles) and fluorescence detection at 330 nm with excitation at 280 nm (downward triangles), respectively (Nölting et al., 1995, 1997a; Nölting, 1998a). At 10°C, the folding rate constants for #1 and #2 are (435 µs)$^{-1}$ and (91 ms)$^{-1}$, respectively, and at 6°C they are (562 µs)$^{-1}$ and (181 ms)$^{-1}$, respectively.

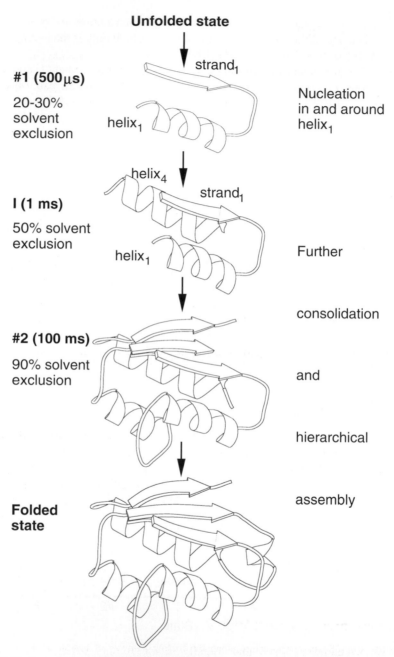

Unfolded state

strand$_1$

#1 (500 μs)

20-30%
solvent
exclusion

helix$_1$

Nucleation
in and around
helix$_1$

helix$_4$

strand$_1$

I (1 ms)

50% solvent
exclusion

helix$_1$

Further

consolidation

#2 (100 ms)

90% solvent
exclusion

and

hierarchical

assembly

**Folded
state**

Fig. 10.20. Structural consolidation along the folding pathway of barstar (Nölting et al., 1995, 1997a; Nölting, 1998a, 2003). #1, I, and #2 are the microsecond transition state, intermediate state, and late transition state, respectively. At 10°C, the intermediate and folded states are populated to 70–80% after 1 ms and 200 ms, respectively. The figure was drawn using the program MOLSCRIPT (Kraulis, 1991). See also Fig. 11.3.

10.8.1
Microsecond transition state

The evidence collected from the kinetic, spectroscopic, equilibrium thermo-dynamic characterization and Φ-value analysis (Sects. 10.6 and 10.7) suggests that in the absence of denaturants, about 500 μs after initiation of the folding reaction, the barstar molecule collapses under roughly 25% solvent exclusion and the polypeptide chain kinks to enable the formation of attractive interactions in the $helix_1$–$loop_1$–$strand_1$ motif (Figs. 10.16, 10.19, 10.20; Nölting et al., 1995, 1997a, b; Nölting, 1998a).

10.8.2
Intermediate

The structure of the early transition state allows the molecule to condense further within a few 100 μs to form an intermediate with strengthened tertiary structure interactions of $helix_1$, $helix_2$, and $helix_4$. Secondary structure interactions in $helix_1$ are further strengthened, and at this stage, $helix_4$ has formed significant amounts of secondary structure too. As judged by the Gibbs free energy, the average structure consolidation of the early intermediate is only 20%. Few, if any, residues form very strong non-native interactions, as suggested by the absence of significantly negative changes in Φ-values (Fig. 10.15). The solvent exclusion has already reached 50%, as judged by the effect of urea on the rate constants of folding and unfolding (Figs. 10.17, 10.19, 10.20; Nölting et al., 1995, 1997a, b).

10.8.3
Late transition state

100 ms later, the still largely fluid-like molecule of small stability has significantly further consolidated and is passing through the slow folding transition state with an average of 50% structure consolidation, as judged by Gibbs free energy, and 90% solvent exclusion, as judged by m-values. This state enables then the rapid formation of the fully folded conformation. Tertiary structure interactions in #2 are detected at many positions in $helix_1$, $helix_2$, $helix_4$, $strand_1$, $strand_2$, and $strand_3$. Strong secondary structure interactions have built up in $helix_1$ and $helix_4$. Nearly complete structure formation is found for $helix_1$ at the positions Ser14 and Leu16 and in $helix_4$ at the position Arg75. The inter-residue contact map suggests that there is little, if any, consolidation of $helix_3$ (Figs. 10.18–10.20; Nölting, 1998a).

10.8.4
Directional propagation of folding

Figs. 10.16 and 10.17 show that initially in the folding reaction, i.e., in #1 ($\approx$500 μs) and I ($\approx$1 ms), amino acid residues that are located close to the middle of the polypeptide chain display little, if any, structure consolidation. In I, mainly $helix_1$ and $helix_4$ have significant Φ-values. In the course of the folding reaction, the consolidation of structure propagates from the termini towards $helix_3$, which not yet formed in #2 (≈ 100 ms). In #2 for most of the residues located close to the middle of the polypeptide chain, the formation of secondary structure is weak, but N- and C-termini are largely consolidated as suggested by Φ-values close to 1. The consolidation of structure in the middle of the sequence is the latest folding event that proceeds mainly between #2 and the fully folded state, F (Figs. 10.18 and 10.20; Nölting, 1998a).

Further, Φ-values close to 1 found for #2 at several positions in $helix_1$ and $helix_4$ suggest a high degree of correct sidechain interlocking at these positions (Figs. 10.14, 10.15, 10.18). Probably, there are still weak non-native interactions in most other parts of the molecule as suggested by Φ-values of less than 1 and by a lower average in Gibbs free energy change of 50% compared with the overall degree of solvent exclusion of 90% (Nölting et al., 1995, 1997a; Nölting, 1998a).

A considerable likelihood of terminal proximity has been predicted for random-flight chains (Flory, 1969; Ptitsyn, 1981), and has been observed in folded structures of proteins (Christopher and Baldwin, 1996). This study (Nölting, 1998a) presents evidence for terminal consolidation in the early stages of folding of barstar (Fig. 10.20). One might speculate that an early consolidation of the N-terminal part of the molecule could represent an advantage in the *in-vivo* protein synthesis. However, the Φ-value analysis of barnase did not reveal this feature (Fersht, 1993) and for CI2 the effect of early terminal consolidation is at the borderline of statistical significance (Itzhaki et al., 1995a; Nölting, 1999), suggesting that this feature of barstar folding is not found in all proteins. A more important aspect of this high resolution of a protein folding pathway is that the discovered nucleation–condensation events (see Sect. 10.11) involve the formation of tertiary structure interactions of structural elements that are distant in the primary structure. A similar behavior has been observed also in CI2 on a lower time scale (Fersht, 1993; Itzhaki et al., 1995a; Nölting, 1999). The obvious advantage of a folding nucleus which is stabilized by interactions of distant structural elements is that it can prevent misfolding more efficiently than a nucleus that would comprise only short-range interactions (Sect. 10.11; Nölting, 1998a).

10.8.5
Cis–trans isomerization

Folded barstar with the *cis*-conformation of the prolyl(48)-peptidyl bond has a significantly higher stability than that with the *trans*-conformation. However, the small, if any, effect of mutation on the ratio of the unfolding rate constant of barstar with *cis*-conformation of the prolyl(48)-peptidyl bond, F_{cis}, versus that for the *trans*-conformation, F_{trans}, (Fig. 10.21) suggests that the *cis–trans* isomerization of this bond is mainly a local effect and has no significant effect on the degree of consolidation of structure in most parts of the molecule (Nölting et al., 1997a; Nölting, 1998a).

Taking the *cis–trans* isomerizations of the prolyl(48)-peptidyl bond into account, the scheme for barstar folding has to be extended according to Fig. 10.22.

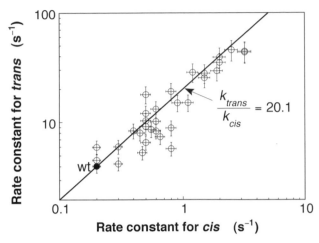

Fig 10.21. Unfolding rate constants for pseudo-wild-type barstar (wt) and several of its mutants with *cis*-conformation of the prolyl(48)-peptidyl bond versus those for the *trans*-conformation in 3 M urea at 11.6°C. The line indicates the ratio of the rate constants which is obtained for pseudo-wild-type. A similar ratio of rate constants is obtained for most of the mutants, which suggests that the *cis–trans* isomerization is mainly a local and not a global folding event.

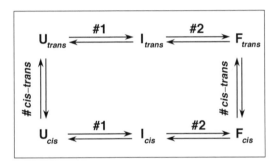

Fig. 10.22. Extended scheme for the folding pathway of barstar. The pathway which leads from the unfolded state with *trans*-conformation of the prolyl(48)-peptidyl bond, U_{trans}, to the folded state with *trans*-conformation, F_{trans}, to the fully folded state with *cis*-conformation, F_{cis}, is the major pathway.

10.8.6
Are there further folding events?

There are no indications of further global folding events with a significant Gibbs free energy contribution: 1. The total of the Φ-value increments calculated using Eq. 8.36 comes to 1, within the experimental error for all mutants, with exception of two mutations located close to Pro48. 2. The high kinetic resolution down to a few μs (Fig. 10.23) suggests that there is no further phase with a fluorescence change of significant amplitude.

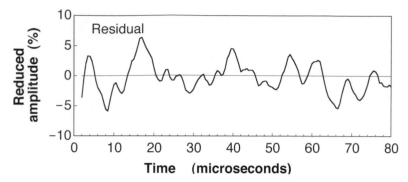

Fig. 10.23. Deviation of the temperature-jump fluorescence trace for the transition $U_{trans} \rightleftarrows I_{trans}$ from the single-exponential curve fit. Only part of the trace is shown; the complete time range for trace and curve fit was 3 ms. There is no evidence for a lag phase preceding the formation of the early intermediate, I_{trans} (Nölting et al., 1997a).

10.9
Structural disorder and misfolding

For a comprehensive Φ-value analysis it is important to distinguish between correctly and wrongly formed interactions. Non-native interactions must exist in the individual stages of sampling of conformations along the folding pathway, in particular in the intermediate and transition states, because otherwise the folding reaction is predicted to take significantly less than 1 ms (Nölting, 1998a, 1999). Intriguingly, few if any significantly negative Φ-value changes are observed (Figs. 10.14 and 10.15), which suggests little, if any, formation of strongly attractive non-native interactions. This finding indicates the absence of significant misfolding on the folding pathway of barstar, i.e., the absence of deep energy traps with slow exchange rates to the native conformation, but rather the presence of structural disorder, i.e., small energy contributions from a large number of interactions. A reason for this behavior may be that the intermediate of barstar corresponds to a relatively shallow valley in the energy-landscape of folding of only 2.5 kJ mol^{-1} (0.6 kcal mol^{-1}) (Nölting et al., 1997a). However, the red-shifted kinetic difference fluorescence spectrum for the fast transition relative to the slow transition (Fig. 10.8), the large CD signal at 222 nm of the intermediate (40% of the folded state), and its large solvent exclusion (50% of the folded state), compared with its stability of only 20% of the stability of the folded state, suggest a significant sidechain disorder in the intermediate (Pfeil, 1993; Pfeil et al., 1993a; Nölting et al., 1995, 1997a; Nölting, 1998a, 1999).

Because for #1 there is no significant cross-correlation between tertiary structure formation in helix$_1$ and helix$_4$ (Fig. 10.16) and since helix$_4$ is still in a disordered state with no stable secondary structure formed yet, no statement about a correct or incorrect tertiary structure alignment can be made for this stage (Nölting, 1998a).

There are, however, indications for a significant degree of correct tertiary structure alignment in the early intermediate: 1. Peptides comprising the $helix_1$–$loop_1$–$helix_2$ motif and $helix_4$, respectively, do not have significant secondary structure under folding conditions (see Sect. 10.10; Nölting et al., 1997a). 2. The inter-residue contact map (Fig. 10.17) shows a strong cross-correlation for secondary and tertiary structure contacts of $helix_1$ and $helix_4$. 3. The average of Φ-values for mainly secondary structure probing and mainly tertiary structure probing mutants is found to be the same (Nölting, 1998a).

The absence of misfolding in the barstar intermediate is in agreement with computer simulations of folding (Wolynes et al., 1995; Nymeyer et al., 1998), which suggest that the so-called glass-transition, i.e., the freezing of conformations into deep energy traps, may occur after formation of most of the native contacts, but is less likely to occur in fluid-like early states (Fig. 10.24; Nölting, 1998a).

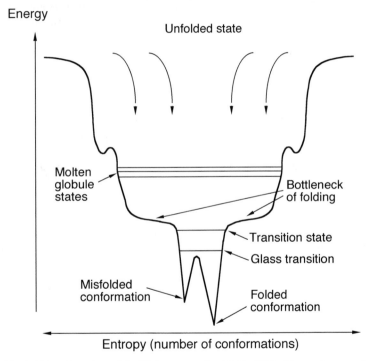

Fig. 10.24. Folding funnel model (Wolynes et al., 1995, 1996; Shoemaker et al., 1997). Computer simulations suggest that early in the folding reaction there is a significant number of weak non-native contacts. The main transition state barrier arises from an entropic bottleneck (Onuchic et al., 1996). After passing through the glass transition, misfolding may occur, i.e., protein molecules may become trapped into conformations that correspond to deep energy valleys from which escape is possible only at a very low rate (Hagen et al., 1995; Shortle et al., 1998). Good folding sequences have energy landscapes where the energy bias towards the fully folded state is larger than the ruggedness of the folding funnel (Nymeyer et al., 1998).

10.10
Structure of peptides of barstar

Four peptides that correspond to sequences in barstar have been synthesized and investigated (Nölting et al., 1997a): 11–29, 33–44, 28–44, and 14–43, that contain the sequences of helix$_1$, helix$_2$, loop$_1$–helix$_2$, and the entire helix$_1$– loop$_1$–helix$_2$ motif, respectively. 40% trifluoroethanol induces helical structure in the 4 peptides, as judged by far-UV circular dichroism (CD). However, in H$_2$O at pH 5.3 and 5°C the two-dimensional ^{1}H-NMR spectra of the peptides 11–29, 33–44, and 28–44 resemble those expected for a random coil state, apart from some specific interactions involving the tryptophan residues 38 and 44 (the 14–43 peptide could not be studied in aqueous solutions because of signal broadness). CD spectra in H$_2$O are also close to the random coil spectrum (Fig. 7.11) with less than 5–10% of the molar ellipticity at 222 nm expected for a helical structure. Less than 5% helical structure is calculated from a titration procedure (Jasanoff and Fersht, 1994). Thus, helix$_1$ and helix$_2$ do not form significant amounts of stable helical structure under folding conditions in the absence of the rest of the protein. The formation of the relatively stable structure in helix$_1$ in the folding intermediate I must, accordingly, be coupled with the formation of long-range interactions that stabilize the helix (Nölting et al., 1997a).

10.11
Nucleation–condensation mechanism of folding

The data presented in Sects. 10.3–10.10 are consistent with a nucleation–condensation model (Abkevich et al., 1994b; Itzhaki et al., 1995a; Fersht, 1995c; Shakhnovich et al., 1996; Shoemaker et al., 1997; Guo and Thirumalai, 1997; Klimov and Thirumalai, 1998; Michnick and Shakhnovich, 1998; Ptitsyn, 1998; Nölting et al., 1997a; Nölting, 1998a, 1999) for folding of barstar in the submillisecond time scale. Although the early stages of barstar folding represent collapsed states, they are clearly not uniformly consolidated. Analogous to the growth of a crystal, a part of the molecule, the so-called nucleation site, forms significantly earlier than other parts of the molecule. In the course of the reaction, the initially diffuse folding nucleus becomes increasingly stabilized as further structure condenses around it. The nucleus of barstar (Fig. 10.20) comprises mainly helix$_1$ and some surrounding structural elements (Nölting et al., 1997a; Nölting, 1998a).

The investigation on barstar was inspired by earlier studies (Matouschek et al., 1989; Fersht et al., 1992; Fersht, 1993; Serrano et al., 1992c) on the ribonuclease barnase that has a clear modular structure (Yanagawa et al., 1993). Indirect methods indicated the occurrence of a distinct early forming folding intermediate, and Φ-value analysis of the later stages of folding suggested a framework mechanism in which a preformed secondary structure of α-helix docked on that of β-sheet (Serrano et al., 1992c). A general scheme has been proposed in which modules of structures in larger globular proteins were formed initially by nucleation–condensation (Abkevich et al., 1994b; Itzhaki et al., 1995a). These modules would then dock, either in a purely stepwise manner or by the processes of docking and nucleation–condensation being coupled, depending on the stabilities of the modules (Itzhaki et al., 1995a). It was predicted, that the more the stabilities of the individual modules, the greater the hierarchical tendency of the folding reaction. However, it was not possible to measure the rate of formation of the folding intermediate of barnase and so there was no direct evidence for an initial nucleation–condensation process (Nölting et al., 1997a).

Application of new T-jump methods (Nölting et al., 1995, 1997a; Nölting, 1996, 1998b), in combination with protein engineering, to the study of early folding events of barstar enabled to test the nucleation–condensation mechanism directly. The results of Φ-value analysis and spectroscopic studies on barstar are directly consistent with a nucleation–condensation model (Fig. 10.25): Peptides that contain helix$_1$ and correspond to parts of a folding nucleus are mainly random under folding conditions in the absence of the rest of the protein, but a nucleus centered around helix$_1$ is substantially formed in the first transition state on the microsecond time scale, and further consolidated in the early-formed intermediate. Many of the rest of the amino acid residues in the protein make weak interactions in the intermediate, which are then more highly consolidated in the later transition state (Figs. 10.16–10.18; Nölting et al., 1997a; Nölting, 1998a).

Considering the astronomically high number of possible conformations of a random polypeptide chain (see Fig. 1.1), proteins may fold with amazingly high rate constants, and folding is an astonishingly efficient process. One of the fastest-folding proteins, a thermostable variant of monomeric lambda repressor, can fold in approximately 20 µs (Burton et al., 1997)! As pointed out in the *Introduction*, folding cannot proceed via random sampling of all possible conformations.

The nucleation–condensation model of protein folding may explain the high speed of folding reactions (Itzhaki et al., 1995a). Nucleation–growth theory predicts that the nuclei of fast-folding protein sequences contain a certain number

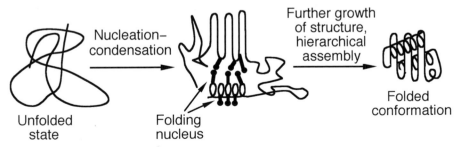

Fig. 10.25. Nucleation–condensation model for protein folding (Abkevich et al., 1994b; Itzhaki et al., 1995a; Fersht, 1995c; Shakhnovich et al., 1996; Nölting et al., 1997a; Nölting, 1998a, 1999). The formation of a few correct secondary and tertiary structure interactions in the folding nucleus catalyzes further folding. The nucleus becomes increasingly stabilized as further structure condenses around it.

of native contacts (Michnick and Shakhnovich, 1998; Mirny et al., 1998). Theoretical studies suggest that the formation of a specific nucleus is a necessary and sufficient condition for subsequent rapid folding to the folded state, that strengthening of interactions in the nucleus is accompanied by acceleration of folding, and that the amino acid residues involved in the nucleus are the most conserved ones within families of evolved sequences (Abkevich et al., 1994b; Shakhnovich et al., 1996; Mirny et al., 1998).

A nucleus which has a (marginal) stability in the presence of some correctly formed tertiary structure interactions is predicted to catalyze folding by preventing misfolding: If secondary structure would form significantly without correct tertiary structure interactions, misalignment of these secondary structure elements could easily occur in proteins which have a complicated chain topology (Nölting, 1998a). This conclusion is consistent (a) with observations on a slower time scale of a nucleation–condensation folding mechanism for chymotrypsin inhibitor 2 (CI2) which suggest that the folding nucleus is not stable in the absence of long-range interactions (Fersht, 1995c; Nölting, 1999), and (b) with observations of the effect of a large number of long-range interactions in the folded state on the speed of folding: Proteins of similar size fold several orders of magnitude faster if they exhibit a relatively simple chain topology with mainly strong local contacts, especially those conductive to helix formation (Viguera et al., 1997; Chan, 1998; Plaxco et al., 1998): Obviously, the more the folded structure is dominated by complex chain topologies with many long-range interactions, the longer it takes to find the conformation of the transition state and the slower the protein will fold.

The nucleation–condensation model is consistent with the folding funnel model (see Fig. 10.24; Wolynes et al., 1995, 1996; Onuchic et al., 1996; Shoemaker et al., 1997). Predictions about the residues involved in the folding nucleus agree for both models (Shoemaker et al., 1997). Fast folding sequences have a low ruggedness of the folding funnel (Nymeyer et al., 1998). Energy landscape theory predicts, in agreement with the experimental results on barstar and CI2 (which contain α- and β-structure), that the folding funnel for small fast-folding α-helical proteins has a transition state roughly half-way to the fully folded state (Onuchic et al., 1996).

For barstar, the speed of the formation of the early intermediate, $\approx (500\,\mu s)^{-1}$, is well below the diffusion limit of folding which is estimated to be around $(1\,\mu s)^{-1}$ for a protein of this size (Hagen et al., 1996), suggesting that a large number of conformational sampling steps takes place, roughly 500–500,000 steps if one assumes $1\,ns - 1\,\mu s$ per step, to funnel the unfolded conformations into the intermediate state. This large number of sampling steps indicates the existence of a significant number of important interactions in the folding nucleus and represents evidence against a diffusion-limited hydrophobic collapse model for this folding step (Nölting et al., 1995, 1997a; Nölting, 1998a; 1999).

11 Highly resolved folding pathways and mechanisms of six proteins

Similar investigations as described for barstar in the previous chapter were done with five further small proteins: barnase, chymotrypsin inhibitor 2 (CI2), the src SH3 domain, Arc repressor, and a tetrameric p53 domain. As detailed in the following sections, it was found that for the main transition states for formation of the native structure of barstar, barnase, CI2, src SH3 domain, and Arc repressor:

1. on average over the molecule, secondary and tertiary structure interactions have built up to the same degree, or at least a high degree, but the built-up of interactions is non-uniformly distributed over the molecule;

2. the most consolidated parts of the molecules form clusters, and these clusters contain a particularly high fraction of amino acid residues that belong to secondary structure elements of the native state;

3. elements of secondary structure have on average a larger relative consolidation than loops, as judged by Φ-values for the main transition states (Nölting and Andert, 2000).

As discussed in the following sections, these observations further reconcile the framework model with the nucleation–condensation mechanism for folding: the amazing speed and efficiency of folding of many proteins can be understood as caused by the catalytic effect of the formation of folding nuclei which comprise significant amounts of tertiary structure interactions, but have a preference for the early formation of regular secondary structure (Nölting and Andert, 2000).

The Φ-values in this chapter are from Matouschek et al., 1992; Serrano et al., 1992a, 1992b; Matouschek and Fersht, 1993; Milla et al., 1995; Itzhaki et al., 1995a; López-Hernández and Serrano, 1996; Nölting et al., 1997a; Nölting, 1998a; Riddle et al., 1999; Mateu et al., 1999; Chiti et al., 1999; Fulton et al., 1999. For reasons of a higher precision, only data from mutants which cause a change of stability $|\Delta\Delta G_{F-U}| > 0.5$ kcal mol^{-1} were used. For more details regarding the study presented in this chapter the reader may also refer to Nölting and Andert (2000).

11.1
General features of the main transition states for the formation of the native structures

As discussed in Chap. 10, different models have been established to explain the surprisingly high speed and efficiency of folding, e.g., the framework model

Models for protein folding

(a) Framework model

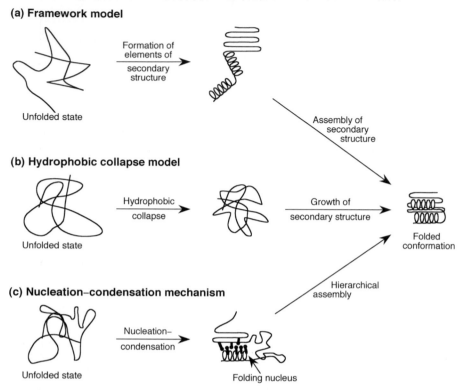

Unfolded state

Formation of
elements of
secondary
structure

Assembly of
secondary
structure

(b) Hydrophobic collapse model

Unfolded state

Hydrophobic
collapse

Growth of
secondary structure

Folded
conformation

(c) Nucleation–condensation mechanism

Hierarchical
assembly

Unfolded state

Nucleation–
condensation

Folding nucleus

Fig. 11.1. Models for protein folding (Nölting and Andert, 2000). **a:** Framework model (Ptitsyn and Rashin, 1975; Kim and Baldwin 1982, 1990; Udgaonkar and Baldwin, 1988). Protein folding is thought to start with the formation of elements of secondary structure. These elements form independently of tertiary structure, or at least before tertiary structure is locked in place. The elements then assemble into the tightly packed native tertiary structure either by diffusion and collision (Karplus and Weaver, 1994) or by propagation of structure in a stepwise manner (Wetlaufer, 1973). **b:** Hydrophobic collapse model for folding (Rackovsky and Scheraga, 1977; Dill, 1985, 1990a, 1990b). The initial event of the folding reaction is thought to be a relatively uniform collapse of the protein molecule, mainly driven by the hydrophobic effect, i.e., the tendency of non-polar groups dissolved in water to cluster together (see Sect 3.4). Stable secondary structure elements can only form in the resulting collapsed state. **c:** Nucleation-condensation mechanism (Fersht 1995c; Itzhaki et al., 1995a; Nölting et al., 1995, 1997a; Shakhnovich et al., 1996; Fersht, 1997, 1999; Kiefhaber et al., 1997; Nölting, 1998a, 1999; Michnick and Shakhnovich, 1998; Otzen and Fersht, 1998): Early formation of a folding nucleus catalyzes further folding. The nucleus is diffuse, but comprises secondary structure interactions and approximately correct tertiary structure interactions (see also Sect. 10.11). This model is consistent with the funnel model (Wolynes et al., 1995; Shoemaker et al., 1999) which focuses on the rapid decrease of the conformational dispersity in the course of the reaction. Some proteins, in particular the ones with larger numbers of amino acid residues, may have several nuclei. The three models may analogously be applied on proteins with multiple transition states on their pathways (not shown). See also Sect. 10.11.

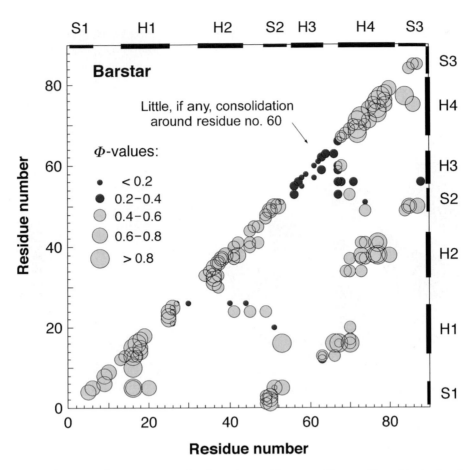

Fig. 11.2. Inter-residue contact map for the main transition state of barstar. The x (abscissa) and y (ordinate) axes indicate the sequence of the protein. Every circle corresponds to a contact of the amino acid residue number x with the residue number y. In this plot, secondary structure contacts are displayed on the diagonal, and tertiary structure contacts are contained in the bulk (see also Fig. 10.18 which shows earlier results on a somewhat different scale of Φ-values). For reasons of simplification, only the bottom right triangle of contacts is shown and the top left triangle with the same information is left out. The sizes and colors of the circles indicate the magnitudes of the Φ-values of the contacts between the residues in the native state that are predicted to be altered by mutation. High Φ-values in the map (large symbols) suggest a high degree of consolidation of structure (about native interaction energies) at the individual positions in the inter-residue contact space. $\Phi \sim 0$ (small symbols) indicates little, if any, formation of stable structure. Φ-values in the range of ~ 0.2–0.8 indicate different probabilities of the consolidation of structure (see Sect. 10.7): for Φ around 0.5 usually only clusters with at least 5 contacts may be used to draw statistically significant conclusions about the presence or absence of a certain degree of structural consolidation. Because of the possibility of non-native interactions, the same region of contacts should be probed by several mutants. Thick bars and the axis labels H1, H2, .. show the positions of helices and thin bars and the axis labels S1, S2, .. show the positions of strands of β-sheets in the native state. For further details see the Chap. 10 and (Nölting, 1998a, 1999, 2003; Nölting and Andert, 2000).

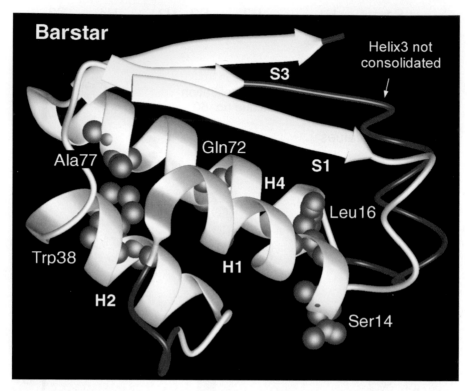

Fig. 11.3. Consolidation of structure in the main transition state of barstar as measured by Φ-value analysis (Nölting and Andert, 2000; Nölting, 2003). The most consolidated secondary structures elements of the molecule are highlighted in yellow, and the parts which are drawn in blue have no fixed structure. Amino acid residues with high Φ-values ($\Phi \geq 0.8$) are highlighted as red spheres.

(Fig. 11.1a), the nucleation–growth mechanism, the diffusion–collision mechanism, the hydrophobic collapse model (Fig. 11.1b), the funnel model (Fig. 10.24), and the nucleation–condensation mechanism (Fig. 11.1c; Sect. 10.11).

As mentioned in Sect. 10.11, the folding nucleus of the nucleation–condensation mechanism is diffuse and mainly consists of several neighboring amino acid residues whose conformations are stabilized by long-range interactions with residues that are remote in sequence. An essential component of this mechanism is the formation of secondary and tertiary structures at the same time (Nölting, 1998a, 1999, 2003; Nölting and Andert, 2000).

The significance of the nucleation–condensation mechanism is that it can make plausible the extreme efficiency of protein folding and can resolve the folding paradox: for example, in a 100-residue protein, the formation of a folding nucleus

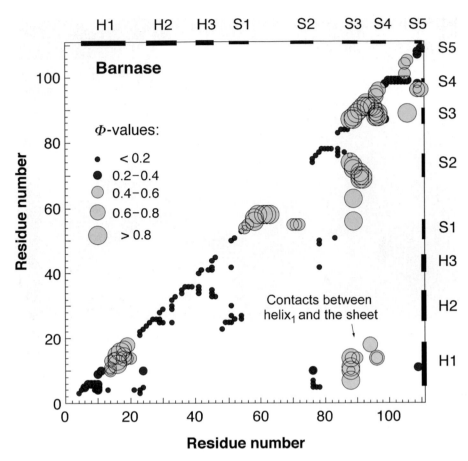

Fig. 11.4. Inter-residue contact map for the main transition state of barnase (Nölting and Andert, 2000; Nölting, 2003). For explanation of the symbols see the legend to Fig. 11.2.

which needs, e.g., 10–20 inter-residue contacts for a sufficient degree of stability probably requires only roughly $10^{10}–10^{20}$ random sampling steps compared to roughly 10^{100} sampling steps for a completely random folding process of the whole molecule (see Chap. 1). Taking into account that in the course of the nucleation the sampling is not completely randomly because steric hindrance and already small energy differences affect the probabilities of different molecular motions, and considering that a sampling step takes roughly 10^{-9} s, one may obtain quite realistic folding times.

A thorough analysis of the structures of the main transition states for the formation of the native state of six proteins by Φ-value analysis presented in the following paragraphs (see also Nölting and Andert, 2000; Nölting, 2003) shows

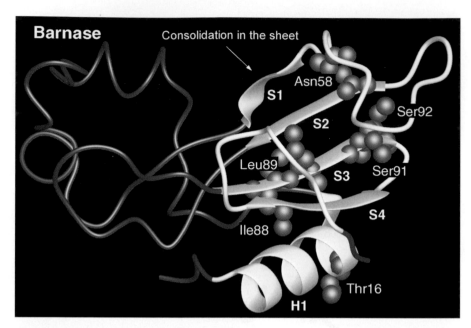

Fig. 11.5. Consolidation of structure in the main transition state of barnase (Nölting and Andert, 2000; Nölting, 2003). Amino acid residues with high Φ-values (Φ ≥ 0.8) are highlighted as red spheres. For further explanation see the legend to Fig. 11.3.

that the nucleation–condensation mechanism is not only valid for a few exceptional proteins, but for at least five of the investigated six proteins, but that their transition states have also some framework-model-like properties. In the following paragraphs, the main transition state structures of the folding pathways of the six proteins are visualized by inter-residue contact maps (Figs. 11.2–11.12, even numbers) and ribbon representations (Figs. 11.3–11.13, odd numbers). This visualization is shown to contribute significantly to a mechanistic understanding of the surprising speed and efficiency of protein folding and to the resolution of the folding paradox (Nölting and Andert, 2000; Nölting, 2003).

The X-ray and NMR measurements show that all six proteins contain considerable amounts of fixed secondary and tertiary structures in their native conformations. Consistent with predictions from the funnel model, the structural consolidation of the main transition states of the four monomeric proteins (barstar, barnase, CI2, src SH3 domain) is about quarter to half way to the native state, as judged by the average of the Φ-values (Tables 1 and 2 in Nölting and Andert, 2000). Accordingly, the average free energy of interactions in the main transition state is about 25–50% of the free energy of interactions in the native state (Nölting and Andert, 2000).

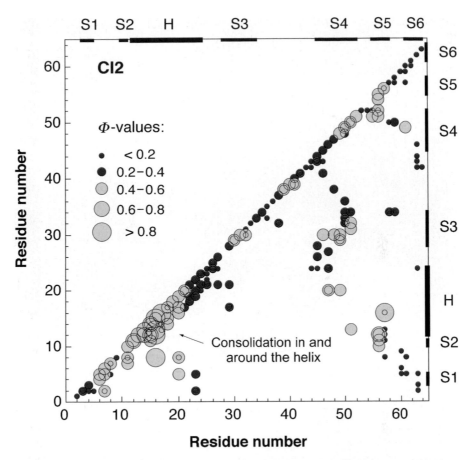

Fig. 11.6. Inter-residue contact map for the main transition state of chymotrypsin inhibitor 2 (CI2) (Nölting and Andert, 2000; Nölting, 2003). For explanation of the symbols see the legend to Fig. 11.2.

Barstar (Figs. 11.2, 11.3) folds via a three-state transition (U $\rightleftarrows$ I $\rightleftarrows$ F, with U, unfolded state; I, intermediate state; F, folded state). Its main transition state (from I to F) has significant amounts of secondary and tertiary structure interactions involving helix$_1$, helix$_2$, helix$_4$, and most parts of the β-sheet, and little, if any, consolidation in helix$_3$ (Nölting and Andert, 2000; Nölting, 2003). For more details see Chap. 10.

Barnase (Figs. 11.4, 11.5) which also folds via a three-state transition (U $\rightleftarrows$ I $\rightleftarrows$ F) clearly shows a cluster of consolidated residues comprising helix$_1$ (Thr16, and slightly lower at His18) and the β-sheet including some of the turns (Asn58 in turn$_3$, Ile88 and Leu89 in strand$_3$, Ser91 and Ser92 in turn$_5$) in its main transition state (from I to F) (Nölting and Andert, 2000; Nölting, 2003).

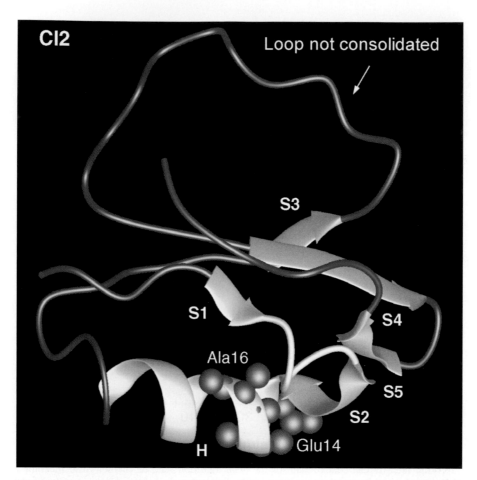

Fig. 11.7. Consolidation of structure in the main transition state of chymotrypsin inhibitor 2 (CI2) (Nölting and Andert, 2000; Nölting, 2003). Amino acid residues with high Φ-values ($\Phi \geq 0.8$) are highlighted as red spheres. For further explanation see the legend to Fig. 11.3.

Also CI2 (Figs. 11.6, 11.7) which has been shown to fold according to a two-state transition (U $\rightleftharpoons$ F) exhibits a clear cluster of consolidated residues which is located in and around the helix. In particular, Glu14 and Ala16 in the helix have Φ-values near 1 (Itzhaki et al., 1995a; Nölting and Andert, 2000; Nölting, 2003). The helix has diffuse contacts with the β-sheet.

A similar clear cluster is observed in the transition state structure of the src SH3 domain (Figs. 11.8, 11.9): the highest Φ-values are found in some residues of strand$_3$ (Ala45, Ser47) and the hairpin (Thr50, Gly51) with connects strand$_3$ with

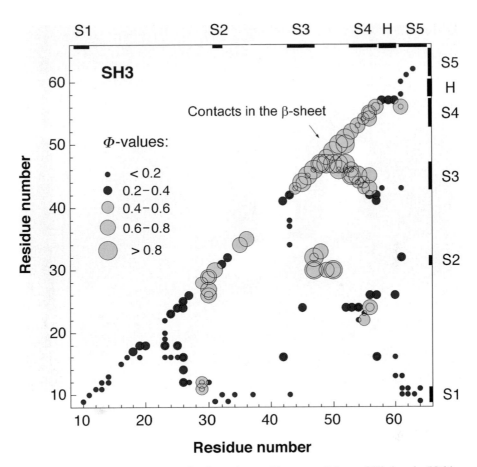

Fig. 11.8. Inter-residue contact map for the main transition state of the src SH3 domain (Nölting and Andert, 2000; Nölting, 2003). For explanation of the symbols see the legend to Fig. 11.2.

strand$_4$. The src SH3 domain also displays two-state folding behavior (U $\rightleftharpoons$ F) (Nölting and Andert, 2000; Nölting, 2003).

Summarizing, all 4 small monomeric proteins display one or two clusters of structural consolidation in some residues which are located in the polypeptide chain about 10 – 30% apart from the N- and C-termini (Figs. 11.2, 11.4, 11.6, 11.8). This position in sequence of the folding nucleus appears not to be a general rule, however, as indicated by observations on other proteins, e.g., acylphosphatase for which the data coverage is not sufficient for a more thorough Φ-value analysis (Nölting and Andert, 2000).

The transition state of the dimeric Arc repressor (Figs. 11.10, 11.11):
1. is in average relatively weakly consolidated,

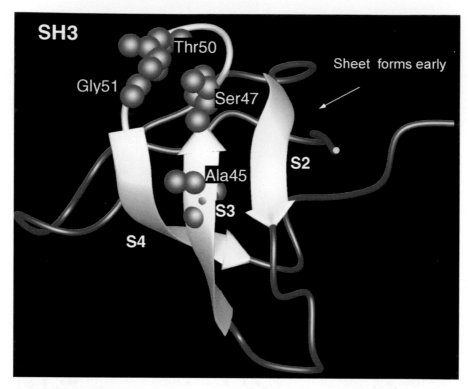

Fig. 11.9. Consolidation of structure in the main transition state of src SH3 domain (Nölting and Andert, 2000; Nölting, 2003). Amino acid residues with high Φ-values ($\Phi \geq 0.8$) are highlighted as red spheres. For further explanation see the legend to Fig. 11.3.

2. has Φ-values larger than 0.4 for only two (Leu19 and Gly30) of the 27 residues probed by mutation with $|\Delta\Delta G_{F-U}| > 0.5$ kcal mol^{-1} (see Table 1 in Nölting and Andert, 2000),
3. has the strongest consolidation near the middle of the sequence (Fig. 11.10),
4. involves a significant number of inter-molecular interactions (Fig. 11.10).
The data show that its transition state structure is affected by both the process of folding and as well the assembly of the monomers (Nölting and Andert, 2000).

In contrast, the main transition state structure of the p53 domain (Figs. 11.12, 11.13) is highly consolidated almost everywhere. The folding model is a four-state transition (Nölting and Andert, 2000):

$$4\,U \rightleftarrows 2\,I'_2 \rightleftarrows 2\,I_2 \rightleftarrows F_4 \; , \qquad (11.1)$$

where U, I'$_2$, I$_2$, and F$_4$ are monomeric unfolded state, first dimeric intermediate state, second dimeric intermediate state, and native tetrameric state, respectively.

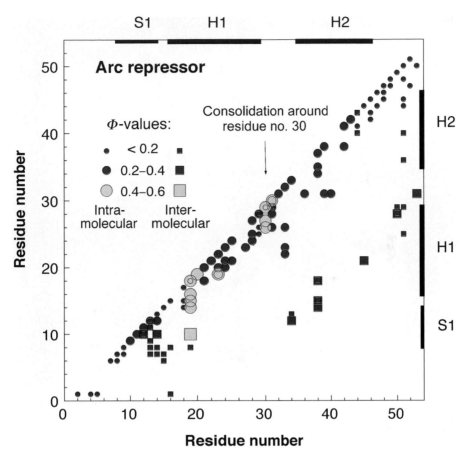

Fig. 11.10. Inter-residue contact map for the main transition state of the dimeric Arc repressor (Nölting and Andert, 2000). For explanation of the symbols see the legend to Fig. 11.2. For Arc repressor there are some quaternary structure contacts because its reaction involves folding and association of the monomers into dimers.

The main transition of this protein is the formation of the tetramer, F_4, from two dimers, $2\ I_2$. Only Φ-values of some mutants which probe interactions at the interface between the two dimers were found to be somewhat lower than 1 which suggests that the interactions between these two dimers are not completely formed (Fig. 11.13). On the other hand, the average of Φ of the transition state for the formation of the early dimers, $2\ I'_2$, from monomers, $4\ U$, is only -0.01 ± 0.03, so the formation of almost all secondary, tertiary, and monomer–monomer quaternary interactions of the molecule occurs in the step $2\ I'_2 \rightleftharpoons 2\ I_2$ (Nölting and Andert, 2000).

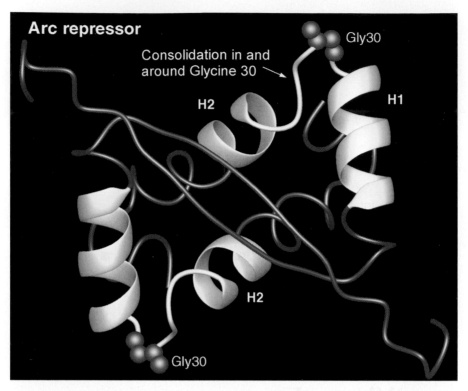

Fig. 11.11. Consolidation of structure in the main transition state of Arc repressor (Nölting and Andert, 2000). For this dimeric protein, folding and association occur nearly concurrently. The atoms of the residue with the largest Φ (0.46) are highlighted as red spheres. For further explanation see the legend to Fig. 11.3.

11.2
Nucleation–condensation

The analysis of the folding behavior of the four monomeric proteins and of the dimeric Arc repressor shows that:

1. their early formed structures are non-uniformly distributed over the molecule (Figs. 11.2–11.11; see also Chap. 10);
2. the most consolidated parts of these structures form one or very few clusters (Figs. 11.2–11.11; see also Chap. 10);
3. these early formed clusters involve secondary and as well tertiary structure interactions (Tables 1–4 in Nölting and Andert, 2000);
4. peptides of barstar (Nölting et al., 1997a; Sect. 10.10), CI2 (Ladurner et al., 1997) and barnase (Neira and Fersht, 1999) have a very low stability;

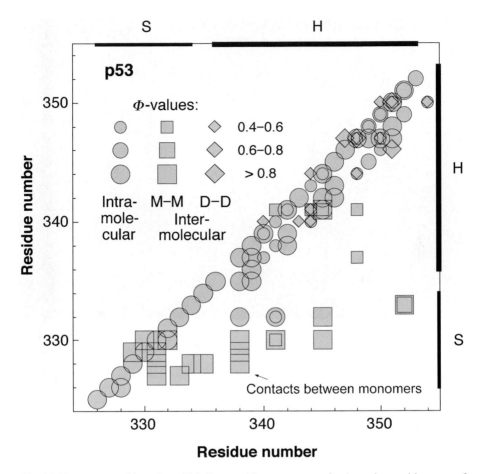

Fig. 11.12. – *continued from Sect. 11.1.* Inter-residue contact map for the main transition state of p53 tetramerization domain (Nölting and Andert, 2000). For explanation of the symbols see the legend to Fig. 11.2. The p53 domain is structurally a dimer of dimers, which involves monomer–monomer (squares) and dimer–dimer (diamonds) contacts.

5. folding involves a significant degree of solvent exclusion in the transition state (Serrano et al., 1992a; Itzhaki et al, 1995a; Milla et al., 1995; Nölting et al., 1997a; Chap. 10).

The combined information suggests that the folding of these five proteins is initiated by nucleation–condensation or a nucleation–condensation-like process (see Sect. 10.11; Nölting and Andert, 2000; Nölting, 2003).

This mechanism has first experimentally been proven for CI2 (Itzhaki et al., 1995a). For the src SH3 domain we see a comparably simple situation with a very clear cluster in and around strand$_3$ and strand$_4$ (Fig. 11.8). In barnase and barstar, which both have an early intermediate on the folding pathway, the cluster of

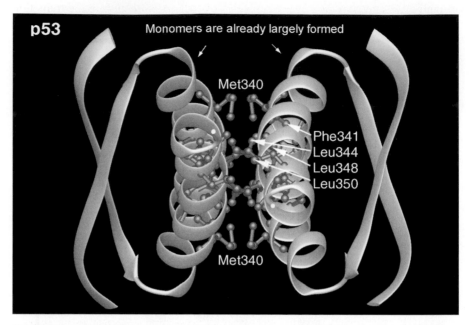

Fig. 11.13. – *continued from Sect. 11.1.* Consolidation of structure in the main transition state of the p53 tetramerization domain (Nölting and Andert, 2000). Residues with $\Phi \leq 0.6$ are highlighted as small red spheres. For this tetrameric protein, the rate-limiting step is the transition from the dimeric to the tetrameric state. The average Φ-value for the early transition state for the formation of the dimers is 0.0.

consolidated residues in the main transition state (Figs. 11.3, 11.5) has already a considerable size. In these two proteins, several nuclei may have docked (Nölting and Andert, 2000). The transition state of Arc repressor is characterized by a relatively low consolidation, but the average degree of formation of inter-molecular interactions ($\Phi = 0.17 \pm 0.02$) is almost the same like that of intra-molecular interactions ($\Phi = 0.21 \pm 0.01$). This indicates that folding of the monomers proceeds almost at the same time or possibly slightly earlier than the docking of the monomers (Milla et al., 1995; Nölting and Andert, 2000; Figs. 11.10, 11.11).

For p53 the mechanism of folding and association appears to be more complicated (Nölting and Andert, 2000), but the Φ-value analysis cannot rule out the possibility of formation of one or several nuclei in the early steps of the reaction, possibly during $2\,\mathrm{I'}_2 \rightleftarrows 2\,\mathrm{I}_2$ (see Sect. 11.1) when most of the contact formation in the molecule proceeds.

One of the interesting observations regarding the folding of proteins is that many folding-related physical parameters have only little dependence on the chain length of the protein. Consistent with this general observation, the nine proteins contained in Table 2 of Nölting and Andert (2000) show only a weak, statistically insignificant, tendency of a larger degree of overall consolidation in the transition state with shorter chain length. This observation is further indication that the folding mechanism involves physical phenomena which largely reduce the exponential growth of possibilities with the number of residues. In particular, also this observation is consistent with a nucleation–condensation mechanism in which the formation of the folding nucleus represents the rate-limiting step of the reaction.

11.3
Framework-model-like properties

As shown in Table 2 in Nölting and Andert, 2000 (for barnase see also Matouschek and Fersht, 1993; Serrano et al., 1992), in five of the six transition states, the average degree of structural consolidation is relatively higher at positions of secondary structure than at positions which form loops in the folded structure. Further, there is an increased percentage of secondary structure forming residues in the most consolidated parts of these five transition states (Table 3 in Nölting and Andert, 2000). These observations support an interpretation towards a partial validity of the framework model (Sect. 11.2; Nölting and Andert, 2000).

On the other hand, with the exception of src SH3 domain, the Φ for tertiary structure contacts is on average similar to Φ for secondary structure contacts (Table 4 in Nölting and Andert, 2000). In the transition state of the src SH3 domain, tertiary structure interactions are weaker than secondary structure interactions, but far from absent. Thus, in all six transition state structures, the degree of tertiary structure consolidation is on average equivalent or almost equivalent to that of secondary structure consolidation. Furthermore, as pointed out in Sect. 11.2, many large peptide fragments of barstar, barnase and CI2 are unstable in the absence of the rest of the molecule. Consequently, the secondary structure elements can gain some degree of stability only in the presence of significantly stabilizing tertiary structure interactions (Nölting and Andert, 2000). So, the picture which emerges is as follows: folding of at least five of the six investigated proteins proceeds via a nucleation–condensation mechanism, but secondary structure interactions appear to be a major driving force in this process.

11.4
Conclusions

According to the presented high resolution of the folding processes (Figs. 11.2–11.13; Tables 1–4 in Nölting and Andert, 2000), nucleation–condensation mechanism and framework model may further be reconciled in a generalized nucleation–condensation mechanism for folding (Nölting and Andert, 2000; Nölting, 2003):

1. Protein folding starts with a collapse of the molecule and structure growth in one or several folding nuclei. These folding nuclei catalyze further folding by reducing the necessary number of sampling steps.
2. The structure of the main transition state for folding consists of one or several folding nuclei which may have docked or may already have attracted further structure around them. This transition state structure contains important tertiary as well as secondary structure interactions, but on average a relatively higher fraction of residues which belong to secondary structure elements than the rest of the molecule.
3. The formation of the structure of the main transition state represents the rate-limiting step of the folding reaction: it enables rapid folding into the native state or a native-like state in which further small structural reorganizations may take place, e.g., prolyl-peptidyl *cis–trans* isomerizations (see Sect. 10.8.5).

It appears that this unified model is valid for the folding of a considerable number of small monomeric and also some dimeric proteins in which additional assembly processes of the monomers take place.

12 Structural determinants of the rate of protein folding

It has been a long-standing question: what makes some proteins fold within less than a millisecond, while others need minutes? This question is directly related to the question what at all makes protein folding so astonishingly fast. It was speculated, that the size or helix content of the protein plays an important role. However, investigations of about 20 small proteins with two-state protein folding kinetics showed only a moderate correlation between helix content and folding speed and even less correlation between molecular weight and folding rate. A breakthrough regarding this question was the discovery that the rate of two-state folding mainly depends on the complexity of the tertiary structure of the protein (Doyle et al., 1997; Chan, 1998; Jackson, 1998; Plaxco et al., 1998; Alm & Baker, 1999; Baker & DeGrado, 1999; Muñoz & Eaton, 1999; Riddle et al., 1999; Baker, 2000; Grantcharova et al., 2000): Proteins with a very complicated tertiary structure fold slower than proteins with a simple tertiary structure. More precisely, the so-called chain topology of the molecule plays a major role for the folding kinetics of small proteins with two-state transitions (Nölting et al., 2003; Nölting, 2003).

12.1
Chain topology as a major structural determinant of two-state folding

The chain topology is a measure of the average distance in sequence of contacts in the molecule. In Fig. 12.1a we see a hypothetical protein molecule with a quite

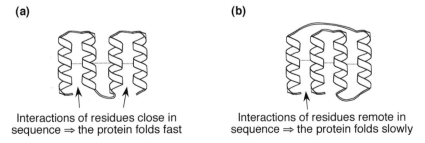

(a)   (b)

Interactions of residues close in   Interactions of residues remote in
sequence ⇒ the protein folds fast   sequence ⇒ the protein folds slowly

Fig. 12.1. Illustration of the effect of the chain topology for a four-helix bundle protein (Nölting et al., 2003). Three contacts in the molecules are shown as dashed lines.

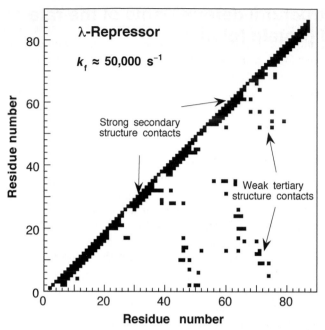

Fig. 12.2. Inter-residue contact map for the folded structure of λ-repressor (Nölting et al., 2003).

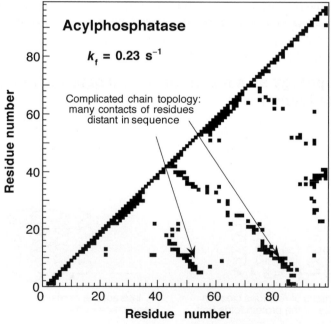

Fig. 12.3. Inter-residue contact map for the folded structure of acylphosphatase (Nölting et al., 2003).

simple chain topology: its tertiary structure contacts are only between amino acid residues relatively close to each other in sequence. However, if the native structure of the molecule would be like shown in Fig. 12.1b, it would have contacts of residues far distant in sequence. This would be the case of a complicated chain topology (Nölting et al., 2003; Nölting, 2003).

This geometrical meaning of chain topology can probably best be visualized in inter-residue contact maps. See, for example, Fig. 12.2 which shows the inter-residue contact map of λ-repressor – an extremely fast folding protein. One can see that this protein has many secondary structure contacts, but only few tertiary structure contacts. Accordingly, its folding rate, k_f, is high: roughly 50,000 s^{-1} for its thermostable variant ! In contrast, acylphosphatase (Fig. 12.3), a protein with many contacts of residues remote in sequence, folds only with $k_f = 0.23$ s^{-1}. So the effect of the chain topology on k_f can be quite dramatic (Nölting et al., 2003).

Originally it was discovered that for proteins with two-state folding kinetics, $-\log(k_f)$ correlates with the so-called contact order, CO. Later a significantly better correlation of $-\log(k_f)$ with the so-called chain topology parameter, CTP, was found (Nölting et al., 2003). CTP is defined as:

$$CTP = \frac{1}{L \cdot N} \sum \Delta S_{i,j}^2 \; , \qquad (12.1)$$

where $\Delta S_{i,j}$ is the separation in sequence between the contacting residues number i and j, L the chain length of the protein molecule, i.e., its number of residues, and N is the total number of inter-residue contacts in the molecule (Nölting et al., 2003).

Fig. 12.4 shows this correlation for 20 small proteins and two peptides. In the range of 10^{-1} $s^{-1} < k_f < 10^8$ s^{-1}, the correlation coefficient is 0.86. The magnitude of CTP slightly varies with the cut-off used in the calculation of the contacts in the

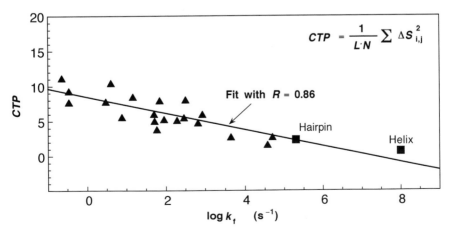

Fig. 12.4. Correlation of CTP with $-\log(k_f)$ for 20 small proteins and two peptides with two-state folding kinetics (Nölting et al., 2003; Nölting, 2003).

protein molecule, but a similarly good correlation between CTP and $-\log (k_f)$ was found for all cut-off distances between 4 and 8.5 Å (Nölting et al., 2003).

12.2
Chain topology of the transition state and implications for the mechanism of folding

The question was raised why the structure of the native state should affect the speed of folding at all? Wouldn't the structure of the transition state be more important? The formation of the transition state structure represents the rate-limiting event in the folding reaction, so its structure should matter and not that of the folded state. This question was answered in Nölting et al. (2003):

1. There is already a good correlation between $-\log (k_f)$ and CTP of the transition state structure.
2. The chain topology of many protein transition states has already some similarities to the chain topology of the native state. In particular, it is likely that the tertiary structure alignment in the transition state is already roughly correct: otherwise the transition would likely lead to a misfolded and not the native state. This is because wrong tertiary structure alignment could not easily be dissolved in the already relatively compact conformations the molecule attains in and after passing through the main transition state.

The latter point is consistent with the nucleation–condensation model in which an essential component of the catalytic action of the folding nucleus is its roughly correct tertiary structure (see Sects. 10.11 and 11.2).

12.3
Further factors

Investigations by Galzitskaya, Finkelstein, and coworkers show that for proteins with three-state kinetics the chain length is an important determinant of folding kinetics (Galzitskaya et al., 2003).

The existence of mutants with largely changed kinetics shows that individual charge interactions can affect the folding rate, k_f, by a factor of 5 and more. In particular, mutations of salt bridges involving lysines or arginines can have a large effect.

$-\text{Log} (k_f)$ correlates also with the number of the residues which belong to β-sheets ($R = 0.73$ for the proteins and peptides in Fig. 12.4; Nölting et al., 2003). This is partially because β-sheets usually involve more long-range contacts than other elements of secondary structure.

Some residues involved in the interlocking of strands of β-sheets participate in the folding nucleus and give a folding-kinetical advantage to members of the sandwich-like protein family (Wilson and Wittung-Stafshede, 2005).

12.4
Ultrafast folding

Using similar methods as described in Chap. 5, in particular temperature jumping (see, e.g., Gruebele et al., 1998; Gruebele, 1999; Hagen and Eaton WA, 2000; Leeson et al., 2000; Mayor et al., 2000; Yamamoto et al., 2000; Hofrichter 2001; Urbanke and Wray, 2001; Callender and Dyer, 2002; Gillespie et al., 2003; Gulotta et al., 2003; Kubelka et al., 2003; Maness et al., 2003; Xu et al., 2003; Arora et al., 2004; Du et al., 2004; Vu et al., 2004; Chung et al., 2005) and optical triggers (see, e.g., Bredenbeck et al., 2005; Buscaglia et al., 2005), several ultrafast-folding proteins were identified. Astonishingly, there are folding rates, k_f, above 100,000 s^{-1}. Examples for ultrafast protein folding events are:

1. the 60-residue three-helix bundle B-domain of protein A from *Staphylococcus aureus* (BdpA): $1/k_f \sim 3$ μs (Arora et al., 2004; Dimitriadis et al., 2004);
2. α_3D, a designed, 73-residue three-helix bundle protein: $1/k_f \sim 3$ μs at ~50°C (Zhu et al., 2003);
3. the 20-residue Trp-cage miniprotein: $1/k_f \sim 4$ μs (Qiu et al., 2002);
4. a 35-residue subdomain of the villin headpiece: $1/k_f \sim 4$ μs at ~27°C (Kubelka et al., 2003);
5. cytochrome b_{562}, a 106-residue four-helix bundle protein: $1/k_f \sim 5$ μs (Wittung-Stafshede et al., 1999);
6. the 61-residue Engrailed homeodomain (En-HD), a three-helix bundle protein: $1/k_f \sim 1$ μs for an intermediate with much native α-helical secondary structure and $1/k_f \sim 25$ μs for folding into the native state (Mayor et al., 2003), probably through a compact native-like transition state (DeMarco et al., 2004). According to simulations, the least stable helix of En-HD, helix$_2$, unfolds in < 450 ps at high temperature (DeMarco et al., 2004).

The formation of monomeric helices takes approximately 100–500 ns (Zhu et al., 2004). Ultrafast folding of proteins is mainly limited by diffusion-collision (Myers and Oas, 2002; Vu et al., 2004), internal friction (Pabit et al., 2004), and the magnitude of the hydrophobic effect (Zhu et al., 2004). Fast formation of helices can drive the ultrafast folding of helical proteins (Vu et al., 2004).

This subclass of proteins behaves differently than the "average" proteins treated in the previous sections: ultrafast-folding proteins have

1. usually a simple chain topology with an accordingly very low chain topology parameter, *CTP* (see Eq. 12.1);
2. very often a small size with a number of amino acid residues of < 80;
3. often a high helix content.

13 Evolutionary computer programming of protein structure and folding

As demonstrated in the previous chapters, protein folding is an extremely complicated, but also highly efficient process. Since a long time it is desired to simulate protein folding on computers. However, one of the major problems to do so is the lack of an efficient mathematical description of multi-body problems like the movements of the structural elements of a biological macromolecule in the course of the folding reaction. It is well known that we do not have an analytical solution for the general three-body problem, i.e., the non-periodical movement of three gravitationally or electromagnetically interacting bodies in space. So, currently it seems hopeless to obtain an analytical solution for the mechanics of a complicated object like a protein. What makes the problem even worse is that humans cannot really foresee in their imagination the non-periodical movement of many bodies in space. This may actually be one of the major reasons why we still have not found a mathematical description of the multi-body problem that can be solved in a simple manner. Currently most physicists believe that the reason for the impossibility of finding an analytical solution for the multi-body problem is the lack of a sufficient number of constants of motion: the equations one can write down contain too many free parameters. But nature can solve with high precision the multi-body problem for macroscopic, non-quantum mechanical bodies. So possibly one reason for our inability to find a simple solution might be that there is something wrong in our mathematics or physics. But so far, nobody could find out the possible shortcoming in our theories, and again this may be because humans currently are not smart enough to map the complicated motions into their brains and then, based on thought experiments, find a comprehensible theoretical description which leads to simple solutions. So what appears to be needed is a machine which is, at least in certain aspects, smarter than humans. But, how can humans develop systems partially more intelligent than themselves? Is it possible, that the ape develops higher mathematics? Probably not.

Fortunately nature shows us the way out of this dilemma: If we cannot directly develop an ultra-capable system, we have to make it evolve itself to a level beyond our direct intellectual abilities. Is this possible? The answer is yes! For example, recently a self-evolving computer program method was created and applied on the calculation of protein structures (Nölting et al., 2004). After a small number of evolution steps, the program is already much more efficient than a program based on rational design (see Fig. 13.1), and intriguingly, it was impossible to say precisely why the evolved structural features of the program code make this program so efficient. The code of the evolved program differs significantly from the original wild-type program, but it appears to be impossible to fully rationalize its higher success.

13.1
Evolution method

The computer evolution method proceeds as follows (Fig. 13.1):

1. An initial, so-called wild-type program is developed. The program contains parts which can be changed (mutated) without causing a complete failure of the program.
2. The following steps are passed through for a certain number of cycles (evolution steps): a) A number of so-called program mutants is created: in the mutants, the changeable parts are altered. b) The performance of the mutants is tested. The best-performing variant within the set of mutants and wild-type program is used as a template for further mutagenesis (step a).
3. The highly evolved program may now be used for the application it was evolved for, but usually also for other, similar tasks.

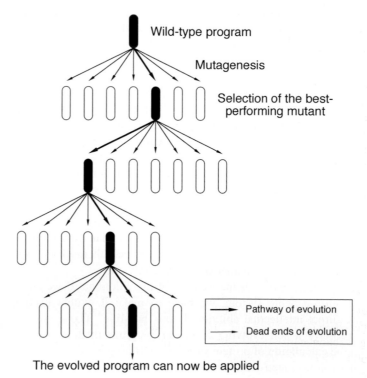

Fig. 13.1. Evolution of a computer program: A wild-type program evolves itself according to an applied evolutionary pressure (see the text and Nölting et al., 2004).

The computer evolution proceeds similar to the evolution of species in nature according to the principles of mutation and selection. Higher evolved programs may also take advantage of the technique of gene shuffling.

In contrast to simple self-learning programs, e.g., programs with neuronal networks, in the course of the evolution, a self-evolving program can significantly change its code and structure: the program evolution is not limited to a certain program structure as it would be the case if, e.g., only parameters in given tables were changed.

As shown in the next section, this evolution method was performed specifically for a program for protein folding and structure predictions (Nölting et al., 2004).

13.2
Protein folding and structure predictions

For the application of self-evolution of a computer program, protein folding and structure prediction is a suitable task because:
1. We do not know enough about the mechanism and mechanics of folding to be able to create, based on rational design, a fast and precise program for folding simulations.
2. The interaction energies are not known precisely enough to determine the conformation with the lowest energy.
3. The number of conformations of unfolded proteins is so astronomically large (see Chap. 1) that it is currently impossible to find the conformation with the lowest energy by calculation of the energies of all conformations even if the interaction energies were precisely known.

For details of the program see Nölting et al. (2004). Briefly, the protein structure is approximated by a hexagonal lattice model (Fig. 13.2). The program was designed to calculate 64-residue proteins. For larger proteins, only the first 64 residues were selected. The potential of interaction between amino acids is V-shaped with a strong repulsion at low distances, an attractive force at moderate distances, and a slowly rising repulsion at larger distances (see Nölting et al., 2004).

Fig. 13.3 shows the structure of the program: The program generates a number of start conformations and tries to fold them. All start conformations consist of random combinations of secondary structure elements. This has been shown to speed up the calculations compared to purely random start conformations. The mutatable parts (genes) of the program encode for the structural representation of the molecule and, most importantly, for the folding reaction (Nölting et al., 2004). Most genes (gene 1 − gene *n* in Fig. 13.3) encode for molecular movements in the folding reaction, e.g., rotations around single bonds, simultaneous rotations around two bonds. Genes of the program can be mutated, e.g., by changing the region of the molecule on which a certain movement is exerted, changing the direction of molecular movement, or by deleting, adding, or exchanging genes.

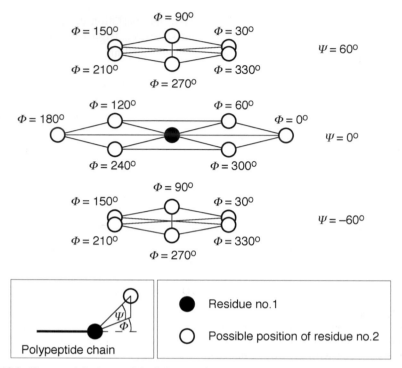

Fig. 13.2. Hexagonal lattice model of the protein structure used for the folding and structure calculations (Nölting et al., 2004). Each amino acid residue is modeled as a sphere. The open circles indicate the possible positions of a residue relative to another one (closed circle). The bond angles Φ and Ψ in the horizontal and vertical direction, respectively, can attain only multiples of $30°$.

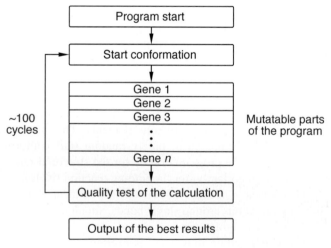

Fig. 13.3. Principle of operation of the self-evolving computer program for protein structure and folding calculations (see the text and Nölting et al., 2004).

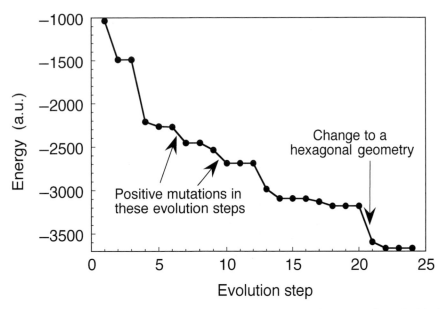

Fig. 13.4. Evolution of the program. In this case, the program evolution was performed with the protein CI2. However, the evolved program has been shown to be applicable also on other small proteins (see, e.g., Nölting et al., 2004). At the evolution step 20, the lattice was changed from the geometry described in (Nölting et al., 2004) to the hexagonal geometry shown in Fig. 13.2.

As shown in Fig. 13.4, the evolution proceeds quite rapidly with some jumps at different stages. The evolutionary pressure applied in this specific case was towards finding deeper minima in the energy landscape within a given time period (Nölting et al., 2004). Other additional pressures are feasible, e.g., towards higher compactness of the protein molecule.

At the cut-off for the contact distance of 8 Å used for all inter-residue contact maps in this section, the native structure of phosphatidylinositol 3-kinase displays extensive clusters of tertiary structure interactions around 26×5 and 56×28 and no interactions around 40×16 in the inter-residue contact space (Fig. 13.5). These important features of the structure were approximately correctly predicted by the simulations (Fig. 13.6). The five structures with the lowest energies consistently display these features in the inter-residue contact maps (not shown).

Analogously, the simulations confirmed important features of the native structure of acyl-coenzyme A binding protein, in particular its completely different structure of inter-residue contact map compared to phosphatidylinositol 3-kinase: in large parts of the inter-residue contact space this protein has only few tertiary structure interactions (Fig. 13.7). In particular, the absence of interactions in the region around 50×10 is correctly predicted (Fig. 13.8).

These are only first results far from perfection, but considering the extreme number of conformations, 10^{64}, it is quite surprising that important features of the protein structures could be calculated by the evolved program which does not use

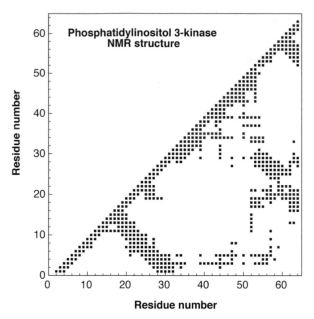

Fig. 13.5. Inter-residue contact map of the NMR structure of phosphatidylinositol 3-kinase (Nölting et al., 2004). The protein has a large number of contacts between the residues around number 26 and around number 5, and between the residues around 56 and around number 28.

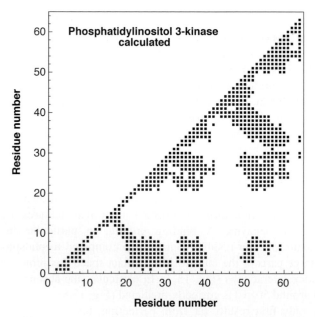

Fig. 13.6. Inter-residue contact map of the calculated structure of phosphatidylinositol 3-kinase. The map reflects the superposition of the three structures with the lowest energy.

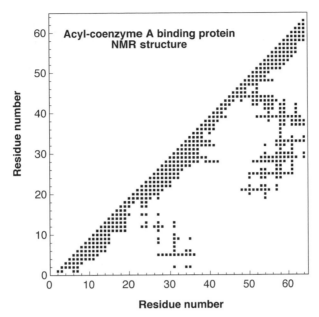

Fig. 13.7. Inter-residue contact map of the NMR structure of the acyl-coenzyme A binding protein.

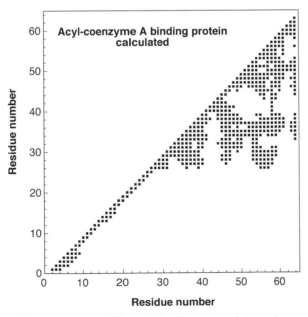

Fig. 13.8. Inter-residue contact map of the calculated structure of the acyl-coenzyme A binding protein. The map reflects the superposition of the three structures with the lowest energy.

any kind of sequence alignment. It should be pointed out that most other successful lattice simulations without sequence alignment were restricted to 27-residue miniproteins which have roughly 10^{37} times fewer conformations. Within the only 20 minutes of simulations on a PC for Figs. 13.6 and 13.8, the evolved program achieved a quality of folding that could not be achieved by the initial program within a 100 times longer period of time, as judged be the energy, compactness and match with known structures of the calculated conformations. This shows that, in principle, it is possible to let a self-evolving program learn how to fold a protein without putting much information about the folding pathway into the initial program. From time to time there are jumps in the evolution (Fig. 13.4), thus, significant further improvement should easily be obtainable by continuing the evolution process and using longer simulations.

13.3
Further potential applications of the evolution method

Obviously, the applicability of the evolution method is by far no means limited to protein folding and structure predictions. Self-evolving programs are probably potentially useful wherever scientific or economical systems are too complex to be rationalized by humans. In fact, it is expected that similar self-evolving systems as presented here will eventually become one of the major sources of further success in science and technology. It appears important that in contrast to many previous approaches to learn from nature (for a survey on evolutionary algorithms see, e.g., Back, 1996), here not only parameters in given tables of a program are changed, but the program is given the ability to change its code. The unprecedented capability of self-evolving systems is that potentially systems and machines can emerge which are not only faster, but truly smarter than humans, at least regarding certain tasks. Potential applications range from the solution of astronomical movements, descriptions of ecosystems, the advanced design of optical devices, robots, and nanomachines, to the prediction of stock indices and sociological and biological phenomena including the evolution of mankind itself.

14 Conclusions

The mystery of submillisecond folding events is in the process of being resolved, along with the development of a variety of new, faster and structurally higher-resolving, kinetic methods. Novel techniques enable the examination of fast folding events on a microsecond to picosecond time scale to be made.

Atomic resolution on a submillisecond time scale is achieved by NMR line broadening experiments if a sufficient population of the species involved is available, and by H/D exchange NMR experiments.

A combination of rapid kinetic-, equilibrium thermodynamic- and protein engineering methods, the so-called Φ-value analysis, enables the structures of folding intermediates at the level of individual amino acid residues, and with microsecond to nanosecond time resolution, to be mapped out. Beyond this, the Φ-value analysis can provide detailed maps for the very important structures of transition states, i.e., for the structures of which formation represents the rate-limiting events in the folding reaction.

A variety of microsecond folding events has been discovered and structurally characterized, ranging from the folding of the B-domain of staphylococcal protein A in ~ 3 μs to the formation of a subdomain in apo-myoglobin in 7 μs at 22°C to the initial collapse of cytochrome c in 40 μs at 40°C to the nucleation–condensation of barstar in 500 μs at 8°C (300 μs at 15°C).

New techniques now enable the predictions of various theoretical models to be tested directly. The folding of the 10-kDa protein barstar and four further small proteins is shown to be consistent with a nucleation–condensation model (Fig. 10.25; Chap. 11). The classical nucleation–growth mechanism invokes the initial formation of a well-defined nucleus followed by growth of structure from it (Wetlaufer, 1973, 1990). The nucleation–condensation mechanism, on the other hand, has a diffuse nucleus that contains a few neighboring amino acid residues whose conformations are stabilized by long-range interactions with amino acid residues that are remote in the primary structure. An essential component of the nucleation–condensation mechanism is that the nucleus and its stabilizing interactions elsewhere in the protein develop concurrently: For example, the α-helices 1, 2, and 4 of barstar are unstable in the absence of long-range interactions, and the rest of the structure is unstable without the interactions with the helices: There is cooperative formation of the folding nucleus and the surrounding structure (Chap. 10; Nölting et al., 1997a; Nölting and Andert, 2000; Nölting, 1998a, 2003).

Many important biological processes and disease states which involve protein folding-, misfolding-, and aggregation reactions have been poorly understood because of lack of extensive information at a molecular level. Examples are spongiform encephalopathies (Creutzfeldt–Jacob disease in humans; BSE in cattle; scrapie in sheep; see Masters and Beyreuter, 1997), Huntington's-, and Alzheimer's diseases.

This book introduces important mathematical, biophysical, and molecular biological methods which can support the high kinetic and structural resolution of such biologically and chemically relevant processes. With these truly interdisciplinary methods, different reaction mechanisms can be far better discriminated than was previously possible. Beyond this, protein structures along the folding pathway can now be visualized at the level of individual amino acid residues in nearly any biologically relevant time scale. This detailed mechanistic knowledge will further aid the understanding of biological processes and disease states, and will eventually help us to find rational ways for re-designing biological processes, and to find cures for diseases.

References

Abkevich VI, Gutin AM, Shakhnovich EI (1994a) Free energy landscape for protein folding kinetics. Intermediates, traps and multiple pathways in theory and lattice model simulations. J. Chem. Phys. 101:6052–6062

Abkevich VI, Gutin AM, Shakhnovich EI (1994b) Specific nucleus as the transition state for protein folding: Evidence from the lattice model. Biochemistry 33:10026–10036

Abkevich VI, Gutin AM, Shakhnovich EI (1996) Improved design of stable and fast-folding model proteins. Folding & Design 1:221–230

Abola EE, Bernstein FC, Bryant SH, Koetzle TF, Weng J (1987) Protein data bank. In: Allen FH, Bergerhoff G, Sievers R (eds) Crystallographic databases – information content, software systems, scientific applications. Data Commission of the International Union of Crystallography, Bonn Cambridge Chester, 107–132.

Abola EE, Sussman JL, Prilusky J, Manning NO (1997) Protein data bank archives of three-dimensional macromolecular structures. Methods Enzymol. 277:556–571.

Agashe VR, Udgaonkar JB (1995) Thermodynamics of denaturation of barstar: Evidence for cold denaturation and evaluation of the interaction with guanidine hydrochloride. Biochemistry 34:3286–3299

Alm E, Baker D (1999) Matching theory and experiment in protein folding. Curr. Opin. Struct. Biol. 9:189–196

Arai M, Kuwajima K (1996) Rapid formation of a molten globule intermediate in refolding of α-lactalbumin. Folding & Design 1:275–287

Arora P, Oas TG, Myers JK (2004) Fast and faster: A designed variant of the B-domain of protein A folds in 3 µsec. Protein Science 13:847–853

Atta-ur-Rahman (1986) Nuclear magnetic resonance. Springer-Verlag, New York Berlin Heidelberg

Ausubel FM, Brent R, Kingston RE, Moore DD, Seidman JG, Smith JA, Struhl K (eds) (1992) Short protocols in molecular biology. Wiley & Sons, New York

Baker D (2000) A surprising simplicity to protein folding. Nature 405:39–42

Baker D, DeGrado WF (1999) Engineering and design. Curr. Opin. Struct. Biol. 9:485–486

Back T (1996) Evolutionary Algorithms in Theory and Practice: Evolution Strategies, Evolutionary Programming, Genetic Algorithms. Oxford University Press

Bacon DJ, Anderson WF (1988) A fast algorithm for rendering space-filling molecule pictures. J. Mol. Graphics 6:219–220

Balbach J, Forge V, Vannuland NAJ, Winder SL, Hore PJ, Dobson CM (1995) Following protein folding in real-time using NMR spectroscopy. Nature Struct. Biol. 2:865–870

Baldwin RL (1996) On-pathway versus off-pathway folding intermediates. Folding & Design 1:R1–R8

Ballew RM, Sabelko J, Gruebele M (1996a) Observation of distinct nanosecond and microsecond protein folding events. Nature Struct. Biol. 3:923–926

Ballew RM, Sabelko J, Reiner C, Gruebele M (1996b) A single-sweep, nanosecond time resolution laser temperature-jump apparatus. Rev. Sci. Instrum. 67:3694–3699

Ballew RM, Sabelko J, Gruebele M (1996c) Direct observation of fast protein folding: The initial collapse of apomyoglobin. Proc. Natl. Acad. Sci. USA 93:5759–5764

Becker OM, Karplus M (1997) The topology of multidimensional potential-energy surfaces: Theory and application to peptide structure and kinetics. J. Chem. Phys. 106:1495–1517

Beyer WH (1991) CRC standard mathematical tables and formulae. CRC Press, Boca Raton Ann Arbor Boston London, 29th Ed.

Billeter M, Schaumann T, Braun W, Wüthrich K (1990) Restrained energy refinement with two different algorithms and force fields of the structure of the α-amylase inhibitor tendamistat determined by NMR in solution. Biopolymers 29:695–706

Bismuto E, Irace G, Sirangelo I, Gratton E (1996) Pressure-induced perturbation of ANS–apomyoglobin complex: Frequency domain fluorescence studies on native and acidic compact states. Prot. Sci. 5:121–126

Björling SC, Goldbeck RA, Paquette SJ, Milder SJ, Kliger DS (1996) Allosteric intermediates in hemoglobin: 1. Nanosecond time-resolved circular dichroism spectroscopy. Biochemistry 35:8619–8627

Bredenbeck J, Helbing J, Kumita JR, Woolley GA, Hamm P (2005) α-helix formation in a photoswitchable peptide tracked from picoseconds to microseconds by time-resolved IR spectroscopy. Proc. Natl. Acad. Sci. USA 102:2379–2384

Booth DR, Sunde M, Bellotti V, Robinson CV, Hutchinson WL, Fraser PE, Hawkins PN, Dobson CM, Radford SE, Pepys MB, Blake CCF (1997) Instability, unfolding and aggregation of human lysozyme variants underlying amyloid fibrillogenesis. Nature 385:787–793

Boxer SG, Anfinrud PA (1994) Casting a cold eye over myoglobin. Nat. Struct. Biol. 1:749–751

Brandl G, Kastner F, Mannschreck A, Nölting B, Andert K, Wetzel R (1991) Chiroptical detection during liquid chromatography. J. Chromatogr. 586:249–254

Bryngelson JD, Onuchic JN, Socci ND, Wolynes PG (1995) Funnels, pathways, and the energy landscape of protein folding: A synthesis. Proteins: Struct. Funct. Genetics 21:167–195

Bundi A, Wüthrich K (1979) ^{1}H NMR parameters of the common amino acids residues measured in aqueous solution of the linear tetrapeptides H–Gly–Gly–X–Ala–OH. Biopolymers 18:285–297

Burova TV, Bernhardt R, Pfeil W (1995) Conformational stability of bovine holo- and apo-adrenodoxin: A scanning calorimetric study. Prot. Sci. 4:909–916

Burton RE, Huang GS, Daugherty MA, Calderone TL, Oas TG (1997) The energy landscape of a fast-folding protein mapped by Ala→Gly substitutions. Nat. Struct. Biol. 4:305–310

Buscaglia M, Kubelka J, Eaton WA, Hofrichter J (2005) Determination of ultrafast protein folding rates from loop formation dynamics. J. Mol. Biol. 2005 347:657–664

Bychkova VE, Ptitsyn OB (1995) Folding intermediates are involved in genetic diseases? FEBS Lett. 359, 6–8

Bycroft M, Matouschek A, Kellis JT, Serrano L, Fersht AR (1990) Detection and characterization of a folding intermediate in barnase by NMR. Nature 346:488–490

Cabani S, Gianni P, Mollica V, Lepori L (1981) Group contributions to the thermodynamic properties of non-ionic organic solutes in dilute aqueous solution. J. Sol. Chem. 10:563–595

Callender R, Dyer RB (2002) Probing protein dynamics using temperature jump relaxation spectroscopy. Curr. Opin. Struct. Biol. 12:628–633

Cantor CR, Schimmel PR (1980) Biophysical chemistry. W. H. Freeman and Co., New York

Cathou RE, Hammes GG (1964) Relaxation spectra of ribonuclease. I. The interaction of ribonuclease with cytidine 3'phosphate. J. Am. Chem. Soc. 86:3240–3245

Cathou RE, Hammes GG (1965) Relaxation spectra of ribonuclease. III. Further investigation of the interaction of ribonuclease and cytidine 3'phosphate. J. Am. Chem. Soc. 87:4674–4680

Causgrove TP, Dyer RB (1993) Protein response to photodissociation of CO from carbonmonoxy myoglobin probed by time-resolved infrared spectroscopy of the amide I band. Biochemistry 32:11985–11991

Chalikian TV, Sarvazyan AP, Breslauer KJ (1994) Hydration and partial compressibility of biological compounds. Biophys. Chem. 51:89–109

Chalikian TV, Gindikin VS, Breslauer KJ (1995) Volumetric characterizations of the native, molten globule and unfolded states of cytochrome c at acidic pH. J. Mol. Biol. 250:291–306

Chalikian TV, Totrov M, Abagyan R, Breslauer KJ (1996) The hydration of globular proteins as derived from volume and compressibility measurements: Cross correlating thermodynamic and structural data. J. Mol. Biol. 260:588–603

Chan HS (1995) Kinetics of protein folding. Nature 373:664–665

Chan HS (1998) Protein folding: Matching speed and locality. Nature 392:761–763

Chan CK, Hofrichter J, Eaton WA (1996) Optical triggers of protein folding. Science 274:628–629

Chan CK, Hu Y, Takahashi S, Rousseau DL, Eaton WA, Hofrichter J (1997) Submillisecond protein folding kinetics studied by ultrarapid mixing. Proc. Natl. Acad. Sci. USA 94:1779–1784

Chen EF, Lapko VN, Song PS, Kliger DS (1997) Dynamics of the N-terminal α-helix unfolding in the photoreversion reaction of phytochrome A. Biochemistry 36:4903–4908

Chen E, Wood MJ, Fink AL, Kliger DS (1998) Time-resolved circular dichroism studies of protein folding intermediates of cytochrome c. Biochemistry 37:5589–5598

Cherepanov AV, De Vries S (2004) Microsecond freeze-hyperquenching: development of a new ultrafast micro-mixing and sampling technology and application to enzyme catalysis. Biochim. Biophys. Acta 1656:1–31

Chiti F, Taddei N, White PM, Bucciantini M, Magherini F, Stefani M, Dobson CM (1999) Mutational analysis of acylphosphatase suggests the importance of topology and contact order in protein folding. Nature Struct. Biol. 6:1005–1009

Chothia C, Hubbard T, Brenner S, Barns H, Murzin A (1997) Protein folds in the all-beta and all-alpha classes. Annu. Rev. Biophys. Biomol. Struct. 26:597–627

Chou PY (1989) Prediction of protein structural class from amino acid compositions. In: Fasman GD (ed) Prediction of protein structure and the principles of protein conformation. Plenum, New York, 549–586

Chou PY, Fasman GD (1977) β-turns in proteins. J. Mol. Biol. 115:135–175

Chou PY, Fasman GD (1978a) Prediction of the secondary structure of proteins from their amino acid sequence. Adv. Enzymol. 47:45–148

Chou PY, Fasman GD (1978b) Empirical predictions of protein conformation. Annu. Rev. Biochem. 47:251–276

Christopher JA, Baldwin TO (1996) Implications of N-terminal and C-terminal proximity for protein folding. J. Mol. Biol. 257:175–187

Chung HS, Khalil M, Smith AW, Ganim Z, Tokmakoff A (2005) Conformational changes during the nanosecond-to-millisecond unfolding of ubiquitin. Proc. Natl. Acad. Sci. USA 102:612–617

Clarke J, Fersht AR (1993) Engineered disulfide bonds as probes of the folding pathway of barnase: Increasing the stability of proteins against the rate of denaturation. Biochemistry 32:4322–4329

Cohn EJ, Edsall JT (1942) Proteins, amino acids, and peptides as ions and dipolar ions. Reinhold, New York

Coligan JE, Dunn BM, Ploegh HL, Speicher DW, Wingfield PT (eds) (1996) Current protocols in protein science. Wiley & Sons, New York

Cowan SW, Schirmer T, Rummel G, Steiert M, Ghosh R, Pauptit RA, Jansonius JN, Rosenbusch JP (1992) Crystal structures explain functional properties of two *E. coli* porins. Nature 358:727–733

Creighton TE (1993) Proteins – structures and molecular properties. W.H.Freeman and Company, New York, 2nd Ed.

Croasmun WR, Carlson MK (eds) (1994) Two-dimensional NMR spectroscopy. Methods in stereochemical analysis. VCH, New York

Cvijovic D, Klinowski J (1995) Taboo search: An approach to the multiple minima problem. Science 267:664–666

Damaschun G, Damaschun H, Gast K, Misselwitz R, Müller JJ, Pfeil W, Zirwer D (1993) Cold denaturation-induced conformational change in phosphoglycerate kinase from yeast. Biochemistry 32:7739–7746

Dauber P, Hagler AT (1980) Crystal packing, hydrogen bonding, and the effect of crystal forces on molecular conformation. Accts. Chem. Res. 13:105–112

Dawson RMC, Elliott DC, Elliott WH, Jones KM (1969) Data for biochemical research. Oxford University Press, 2nd Ed.,

DeMaeyer LCM (1969) Electric field methods. Methods Enzymol. 16:80–118

DeMarco ML, Alonso DOV, Daggett V (2004) Diffusing and colliding: The atomic level folding/unfolding pathway of a small helical protein. J. Mol. Biol. 341:1109–1124

Dill KA (1985) Theory for the folding and stability of globular proteins. Biochemistry 24:1501–1509

Dill KA (1990a) Dominant forces in protein folding. Biochemistry 29:7133–7155

Dill KA (1990b) The meaning of hydrophobicity. Science 250:297–298

Dill KA, Chan HS (1997) From Levinthal to pathways to funnels. Nature Struct. Biol. 4:10–19

Dimitriadis G, Drysdale A, Myers JK, Arora P, Radford SE, Oas TG, Smith DA (2004) Microsecond folding dynamics of the F13W G29A mutant of the B domain of staphylococcal protein A by laser-induced temperature jump. Proc. Natl. Acad. Sci. USA 101:3809–3814

Dolgikh DA, Gilmanshin RI, Brazhnikov EV, Bychkova VE, Semisotnov GV, Venjaminov SY, Ptitsyn OB (1981) α-Lactalbumin – compact state with fluctuating secondary structure. FEBS Lett. 136:311–315

Doyle R, Simons K, Qian H, Baker D (1997) Local interactions and the optimization of protein folding. Proteins 29:282–291

Du D, Zhu Y, Huang CY, Gai F (2004) Understanding the key factors that control the rate of β-hairpin folding. Proc. Natl. Acad. Sci. USA 101:15915–15920

Duddeck H (1995) Günther Snatzke. Liebigs Annalen 6:I–XIII

Dyer RB, Einarsdóttir O, Killough PM, López-Garriga JJ, Woodruff WH (1989) Transient binding of photodissociated CO to CuB+ of eukaryotic cytochrome oxydase at ambient temperature. Direct evidence from time-resolved infrared spectroscopy. J. Am. Chem. Soc. 111:7657–7659

Dyer RB, Peterson KA, Stoutland PO, Woodruff WH (1994) Picosecond infrared study of the photodynamics of carbonmonoxy cytochrome c oxydase. Biochemistry 33:500–507

Dyson HJ, Wright PE (1996) Insights into protein folding from NMR. Annu. Rev. Phys. Chem. 47:369–395

Eaton WA, Thompson PA, Chan CK, Hagen SJ, Hofrichter J (1996a) Fast events in protein folding. Structure 4:1133–1139

Eaton WA, Henry ER, Hofrichter J (1996b) Nanosecond crystallographic snapshots of protein structural changes. Science 274:1631–1632

Eaton WA, Muñoz V, Thompson PA, Chan CK, Hofrichter J (1997) Submillisecond kinetics of protein folding. Curr. Opin. Struct. Biol. 7:10–14

Eftink MR, Shastry MCR (1997) Fluorescence methods for studying kinetics of protein folding reactions. Methods Enzymol. 278:258–286

Eggers F, Kustin K (1969) Ultrasonic methods. Methods Enzymol. 16:55–80

Eigen M (1996) Prionics or the kinetic basis of prion diseases. Biophys. Chem. 63:A1–A18

Eigen M, deMaeyer L (1963) In: Friess SL, Lewis ES and Weissberger A (eds) Techniques of Organic Chemistry. Wiley Interscience, New York, pp. 895–1054

Eigen M, Hammes GG, Kustin K (1960) Fast reactions of imidazole studies with relaxation spectroscopy. J. Am. Chem. Soc. 82:3482–3483

Einterz CM, Lewis JW, Milder SJ, Kliger DS (1985) Birefringence effects in transient circular dichroism measurements with application to the photolysis of carbonmonoxy hemoglobin and carbonmonoxy myoglobin. J. Phys. Chem. 89:3845–3853

Eliezer D, Yao J, Dyson HJ, Wright PE (1998) Structural and dynamic characterization of partially folded states of apomyoglobin and implications for protein folding.

Nat. Struct. Biol. 5:148–155

Elöve GA, Chaffotte AF, Roder H, Goldberg ME (1992) Early steps in cytochrome c folding probed by time-resolved circular dichroism and fluorescence spectroscopy. Biochemistry 31:6876–6883

Elöve GA, Bhuyan AK, Roder H (1994) Kinetic mechanism of cytochrome c folding: Involvement of the heme and its ligands. Biochemistry 33:6925–6935

Englander SW, Mayne L (1992) Protein folding studied using hydrogen exchange labeling and two-dimensional NMR. Annu. Rev. Biophys. Biomol. Struct. 21:243–265

Ernst RR, Anderson WA (1966) Application of Fourier transform spectroscopy to magnetic resonance. Rev. Sci. Instrum. 37:93–103

Esquerra RM, Lewis JW, Kliger DS (1997) An improved linear retarder for time-resolved circular dichroism studies. Rev. Sci. Instrum. 68:1372–1376

Feltch SM, Stuehr JE (1979) Relaxation studies of enzymes: Rapid isomerization in desoxyribonuclease I. Biochemistry 18:2000–2004

Fersht AR (1985) Enzyme structure and mechanism. Freeman and company, New York, 2nd Ed.

Fersht AR (1992) Pathway of protein folding. Faraday Discussions 93:183–193

Fersht AR (1993) Protein folding and stability: The pathway of folding of barnase. FEBS Lett. 325:5–16

Fersht AR (1995a) Characterizing transition states in protein folding: An essential step in the puzzle. Curr. Opin. Struct. Biol. 5:79–84

Fersht AR (1995b) Mapping the structures of transition states and intermediates in folding: Delineation of pathways at high resolution. Phil. Transactions Royal Soc. London B: Biol. Sci. 348:11–15

Fersht AR (1995c) Optimization of rates of protein folding: The nucleation–condensation mechanism and its implications. Proc. Natl. Acad. Sci. USA 92:10869–10873

Fersht AR (1997) Nucleation mechanisms in protein folding. Cur. Opin. Struct. Biol. 7:3–9

Fersht AR (1999) Structure and mechanism in protein science. New York. WH Freeman and Co.

Fersht AR, Bycroft M, Horovitz A, Kellis JT, Matouschek A, Serrano L (1991) Pathway and stability of protein folding. Phil. Transactions Royal Soc. London B: Biol. Sci. 332:171–176

Fersht AR, Matouschek A, Serrano L (1992) The folding of an enzyme. 1. Theory of protein engineering analysis of stability and pathway of protein folding. J. Mol. Biol. 224:771–782

Fersht AR, Itzhaki LS, ElMasry NF, Matthews JM, Otzen DE (1994) Single versus parallel pathways of protein folding and fractional formation of structure in the transition state. Proc. Natl. Acad. Sci. USA 91:10426–10429

Fink AL (1995) Compact intermediate states in protein folding. Annu. Rev. Biophys. Biomol. Struct. 24:495–522

Fink AL, Oberg KA, Seshadri S (1998) Discrete intermediates versus molten globule models for protein folding: Characterization of partially folded intermediates of apomyoglobin. Folding & Design 3:19–25

Finkelstein AV (1997) Protein structure: What is it possible to predict now? Curr. Opin. Struct. Biol. 7:60–71

Finkelstein AV, Badretdinov AY (1997) Rate of protein folding near the point of thermodynamic equilibrium between the coil and the most stable chain fold. Folding & Design 2:115–121

Fisher MT, Sligar SG (1987) Temperature-jump relaxation kinetics of the P-450cam spin equilibrium. Biochemistry 26:4797–4803

Flory PJ (1969) Statistical mechanics of chain molecules. Wiley, New York

Foguel D, Weber G (1995) Pressure-induced dissociation and denaturation of allophycocyanin at subzero temperatures. J. Biol. Chem. 270:28759–28766

Franzen S, Bohn B, Poyart C, Martin JL (1995) Evidence for sub-picosecond heme doming in hemoglobin and myoglobin: A time-resolved resonance Raman comparison of carbonmonoxy and deoxy species. Biochemistry 34:1224–1237

French TC, Hammes GG (1969) The temperature-jump method. Methods Enzymol. 16:3–30

Freund SMV, Wong KB, Fersht AR (1996) Initiation sites of protein folding by NMR analysis. Proc. Natl. Acad. Sci. USA 93:10600–10603

Frieden C, Hoeltzli SD, Ropson IJ (1993) NMR and protein folding: Equilibrium and stopped-flow studies. Protein Science 2:2007–2014

Fulton KF, Main ERG, Daggett V, Jackson SE (1999) Mapping the interactions present in the transition state for unfolding/folding of FKBP12. J. Mol. Biol. 291:445–461

Galzitskaya OV, Garbuzynskiy SO, Ivankov DN, Finkelstein AV (2003) Chain length is the main determinant of the folding rate for proteins with three-state folding kinetics. Proteins: Struct. Funct. Genetics 51:162–166

Gast K, Damaschun G, Damaschun H, Misselwitz R, Zirwer D (1993) Cold denaturation of yeast phosphoglycerate kinase: Kinetics of changes in secondary structure and compactness on unfolding and refolding. Biochemistry 32:7747–7752

Gast K, Damaschun G, Desmadril M, Minard P, Müller-Frohne M, Pfeil W, Zirwer D (1995) Cold denaturation of yeast phosphoglycerate kinase: Which domain is more stable. FEBS Lett. 358:247–250

Gast K, Noppert A, Müller-Frohne M, Zirwer D, Damaschun G (1997) Stopped-flow dynamic light scattering as a method to monitor compaction during protein folding. Eur. Biophys. J. Biophys. Lett. 25:211–219

Gavish B, Gratton E, Hardy CJ (1983a) Adiabatic compressibility of globular proteins. Proc. Natl. Acad. Sci. USA 80:750–754

Gavish B, Gratton E, Hardy CJ, Stdenis A (1983b) Differential sound velocity apparatus for the measurement of protein solutions. Rev. Sci. Instrum. 54:1756–1760

Gillespie B, Vu DM, Shah PS, Marshall SA, Dyer RB, Mayo SL, Plaxco KW (2003) NMR and temperature-jump measurements of de novo designed proteins demonstrate rapid folding in the absence of explicit selection for kinetics. J. Mol. Biol. 330:813–819

Gilmanshin R, Williams S, Callender RH, Woodruff WH, Dyer RB (1997a) Fast events in protein folding: Relaxation dynamics and structure of the I-form of apomyoglobin. Biochemistry 36:15006–15012

Gilmanshin R, Williams S, Callender RH, Woodruff WH, Dyer RB (1997b) Fast events in protein folding: Relaxation dynamics of secondary and tertiary structure in native apomyoglobin. Proc. Natl. Acad. Sci. USA 94:3709–3713

Gilmanshin R, Callender RH, Dyer RB (1998) The core of apomyoglobin E-form folds at the diffusion limit. Nat. Struct. Biol. 5:363–365

Goldbeck RA (1988) Sign variation in the magnetic circular dichroism spectra of π-substituted porphyrins. Accts. Chem. Res. 21:95–101

Goldbeck RA, Kliger DS (1993) Nanosecond time-resolved absorption and polarization dichroism spectroscopies. Methods Enzymol. 226:147–177

Goldbeck RA, Dawes TD, Einarsdóttir O, Woodruff WH, Kliger DS (1991) Time-resolved magnetic circular dichroism spectroscopy of photolyzed carbonmonoxy cytochrome c oxidase (cytochrome aa3). Biophys. J. 60:125–134

Goldbeck RA, Einarsdóttir O, Dawes TD, O'Connor DB, Surerus KK, Fee JA, Kliger DS (1992) Magnetic circular dichroism study of cytochrome ba3 from *Thermus thermophilus*: Spectral contributions from cytochrome b and a3 and nanosecond spectroscopy of CO photodissociation intermediates. Biochemistry 31:9376–9387

Goldenberg DP, Frieden RW, Haack JA, Morrison TB (1989) Mutational analysis of a protein-folding pathway. Nature 338:127–133

Goloubinoff P, Gatenby AA, Lorimer GH (1989a) GroE heat shock proteins assist assembly of foreign prokaryotic ribulose bisphosphate carboxylase oligomers in *Escherichia coli*. Nature 337:44–47

Goloubinoff P, Christeller JT, Gatenby AA, Lorimer GH (1989b) Reconstruction of active dimeric ribulose bisphosphate carboxylase from an unfolded state depends on 2 chaperonin proteins and Mg–ATP. Nature 342:884–888

Grantcharova VP, Riddle DS, Baker D (2000) Long-range order in the src SH3 folding transition state. Proc. Natl. Acad. Sci. USA 97:7084–7089

Griebenow K, Klibanov AM (1997) Can conformational changes be responsible for solvent and excipient effects on the catalytic behavior of subtilisin Carlsberg in organic solvents? Biotechnol. Bioeng. 53:351–362

Griko YV, Privalov PL (1992) Calorimetric study of the heat and cold denaturation of β-lactoglobulin. Biochemistry 31:8810–8815

Gross M, Jaenicke R (1994) Proteins under pressure: The influence of high hydrostatic pressure on structure, function and assembly of proteins and protein complexes. Eur. J. Biochem. 221:617–630

Gruebele M (1999) The fast protein folding problem. Annu. Rev. Phys. Chem. 50:485–516

Gruebele M, Sabelko J, Ballew RM, Ervin J (1998) Laser temperature jump induced protein refolding, Acc. Chem. Res. 31:699–707

Gulotta M, Rogatsky E, Callender RH, Dyer RB (2003) Primary folding dynamics of sperm whale apomyoglobin: core formation. Biophys J. 84:1909–1918

Guo Z, Thirumalai D (1997) The nucleation–collapse mechanism in protein folding: Evidence for the non-uniqueness of the folding nucleus. Folding & Design 2:377–391

Gussakovsky EE, Haas E (1995) 2 Steps in the transition between the native and acid states of bovine α-lactalbumin detected by circular polarization of luminescence: Evidence for a premolten globule state. Prot. Science 4:2319–2326

Hagen SJ, Eaton WA (2000) Two-state expansion and collapse of a polypeptide. J. Mol. Biol. 301:1019–1027

Hagen SJ, Hofrichter J, Eaton WA (1995) Protein reaction kinetics in a room-temperature glass. Science 269:959–962

Hagen SJ, Hofrichter J, Szabo A, Eaton WA (1996) Diffusion-limited contact formation in unfolded cytochrome c: Estimating the maximum rate of protein folding. Proc. Natl. Acad. Sci. USA 93:11615–11617

Hagen SJ, Hofrichter J, Eaton WA (1997) The rate of intrachain diffusion of unfolded cytochrome c. J. Phys. Chem. B 101:2352–2365

Hagler AT, Dauber P, Lifson S (1979) Consistent force field studies of intermolecular forces in hydrogen bonded crystals. III. The C=O···H–O hydrogen bond and the analysis of the energetics and packing of carboxylic acids. J. Am. Chem. Soc. 101:5131–5141

Hammes GG, Roberts PB (1969) Dynamics of helix-coil transition in poly-L-ornithine. J. Am. Chem. Soc. 91:1812–1816

Harper ET, Rose GD (1993) Helix stop signals in proteins and peptides: The capping box. Biochemistry 32:7605–7609

Hartley RW (1988) Barnase and barstar: Expression of its cloned inhibitor permits expression of a cloned ribonuclease. J. Mol. Biol. 202:913–915

Hoeltzli SD, Frieden C (1995) Stopped-flow NMR spectroscopy: Real-time unfolding studies of 6-^{19}F tryptophan-labeled *Escherichia coli* dihydrofolate reductase. Proc. Natl. Acad. Sci. USA 92:9318–9322

Hoeltzli SD, Frieden C (1996) Real-time refolding studies of 6-^{19}F-tryptophan-labeled *E. coli* dihydrofolate reductase using stopped-flow NMR spectroscopy. Biochemistry 35:16843–16851

Hoeltzli SD, Frieden C (1998) Refolding of 6-^{19}F-tryptophan-labeled *E. coli* dihydrofolate reductase in the presence of ligand: A stopped-flow NMR spectroscopy study. Biochemistry 37:387–398

Hofrichter J (2001) Laser temperature-jump methods for studying folding dynamics. Methods Mol. Biol. 168:159–191

Hofrichter J, Henry ER, Szabo A, Murray LP, Ansari A, Jones CM, Coletta M, Falconi G, Brunori M, Eaton WA (1991) Dynamics of the quaternary conformational change in trout hemoglobin. Biochemistry 30:6583–6598

Hu XH, Frei H, Spiro TG (1996) Nanosecond step-scan FTIR spectroscopy of hemoglobin: Ligand recombination and protein conformational changes. Biochemistry 35:13001–13005

Huang GS, Oas TG (1995) Submillisecond folding of monomeric λ-repressor. Proc. Natl. Acad. Sci. USA 92:6878–6882

Hubbard T, Tramontano A (1996) Update on protein structure prediction: Results of the 1995 IRBM workshop. Folding & Design 1:R55–R63

Hunt JF, Weaver AJ, Landry SJ, Gierasch L, Deisenhofer J (1996) The crystal structure of the GroES co-chaperonin at 2.8 Å resolution. Nature 379:37–45

Itzhaki LS, Evans PA, Dobson CM, Radford SE (1994) Tertiary interactions in the folding pathway of hen lysozyme: Kinetic studies using fluorescent probes. Biochemistry 33:5212–5220

Itzhaki LS, Otzen DE, Fersht AR (1995a) The structure of the transition state for folding of chymotrypsin inhibitor 2 analyzed by protein engineering methods: Evidence for a nucleation–condensation mechanism for protein folding. J. Mol. Biol. 254:260–288

Itzhaki LS, Neira JL, Ruiz-Sanz J, PratGay Gde, Fersht AR (1995b) Search for nucleation sites in smaller fragments of chymotrypsin inhibitor 2. J. Mol. Biol. 254:289–304

Jackson SE (1998) How do small single-domain proteins fold? Fold. Des. 3:R81–R91

Jasanoff A, Fersht AR (1994) Quantitative determination of helical propensities from trifluoroethanol titration curves. Biochemistry 33:2129–2135

Jentoft JE, Neet E, Stuehr JE (1977) Relaxation spectra of yeast hexokinases. Isomerization of the enzyme. Biochemistry 16:117–121

Johnson WC Jr (1990) Protein secondary structure and circular dichroism: A practical guide. Proteins: Struct. Funct. Genetics 7:205–214

Jones CM, Henry ER, Hu Y, Chan CK, Luck SD, Bhuyan A, Roder H, Hofrichter J, Eaton WA (1993) Fast events in protein folding initiated by nanosecond laser photolysis. Proc. Natl. Acad. Sci. USA 90:11860–11864

Jung C, Hoa GHB, Davydov D, Gill E, Heremans K (1995) Compressibility of the heme pocket of substrate-analog complexes of cytochrome P450cam-CO: The effect of hydrostatic pressure on the Soret band. Eur. J. Biochem. 233:600–606

Jung C, Ristau O, Schulze H, Sligar SG (1996) The CO stretching mode infrared spectrum of substrate-free cytochrome P450cam-CO: The effect of solvent conditions, temperature, and pressure. Eur. J. Biochem. 235:660–669

Kalnin NN, Kuwajima K (1995) Kinetic folding and unfolding of staphylococcal nuclease and its six mutants studied by stopped-flow circular dichroism. Proteins: Struct. Funct. Genetics 23:163–176

Karplus M, Sali A (1995) Theoretical studies of protein folding and unfolding. Curr. Opin. Struct. Biol. 5:58–73

Karplus M, Weaver DL (1994) Protein folding dynamics: the diffusion–collision model and experimental data. Protein Science 3:650–668.

Katayanagi K, Miyagawa M, Matsushima M, Ishikawa M, Kanaya S, Nakamura H, Ikehara M, Matsuzaki T, Morikawa K (1992) Structural details of ribonuclease H from *Escherichia coli* as refined to an atomic resolution. J. Mol. Biol. 223:1029–1052

Kharakoz DP, Bychkova VE (1997) Molten globule of human α-lactalbumin: Hydration, density, and compressibility of the interior. Biochemistry 36:1882–1890

Kharakoz DP, Sarvazyan AP (1993) Hydrational and intrinsic compressibilities of globular proteins. Biopolymers 33:11–26

Khorasanizadeh S, Peters ID, Butt TR, Roder H (1993) Folding and stability of a tryptophan-containing mutant of ubiquitin. Biochemistry 32:7054–7063

Kiefhaber T, Baldwin RL (1995) Intrinsic stability of individual α-helices modulates structure and stability of the apomyoglobin molten globule form. J. Mol. Biol. 252:122–132

Kiefhaber T, Labhardt AM, Baldwin RL (1995) Direct NMR evidence for an intermediate preceding the rate-limiting step in the unfolding of ribonuclease A. Nature 375:513–515

Kiefhaber T, Bachmann A, Wildegger G, Wagner C (1997) Direct measurement of nucleation and growth rates in lysozyme folding. Biochemistry 36:5108–5112

Kim PS, Baldwin RL (1982) Specific intermediates in the folding reactions of small proteins and the mechanism of protein folding. Annu. Rev. Biochem. 51:459–489

Kim PS, Baldwin RL (1990) Intermediates in the folding reactions of small proteins. Annu. Rev. Biochem. 59:631–660

Kim Y, Grable JC, Love R, Greene P, Rosenberg JM (1990) Refinement of EcoRI endonuclease crystal structure: A revised protein chain tracing. Science 249:1307–1309

Klibanov AM (1989) Enzymatic catalysis in anhydrous organic solvents. Trends Biochem. Sci. 14:141–144

Klibanov AM (1997) Why are enzymes less active in organic solvents than in water? Trends Biotechnol. 15:97–101

Klimov DK, Thirumalai D (1998) Lattice models for proteins reveal multiple folding nuclei for nucleation–collapse model. J. Mol. Biol. 282:471–492

Korolev S, Nayal M, Barnes WM, DiCera E, Waksman G (1995) Crystal structure of the large fragment of *Thermus aquaticus* DNA polymerase I at 2.5 Å resolution: Structural basis for thermostability. Proc. Natl. Acad. Sci. USA 92:9264–9268

Kraulis PJ (1991) MOLSCRIPT: A program to produce both detailed and schematic plots of protein structures. J. Appl. Cryst. 24:946–950

Kubelka J, Eaton WA, Hofrichter J (2003) Experimental tests of villin subdomain folding simulations. J. Mol. Biol. 329:625–630

Kumaraswamy VS, Lindley PF, Slingsby C, Glover ID (1996) An eye lens protein–water structure: 1.2 Å resolution structure of γB crystallin at 150K. Acta Cryst. D 52:611–622

Kunugi S, Suzuki N, Nishimoto S, Morisawa T, Yoshida M (1997) Kinetic study of carboxypeptidase Y catalyzed peptide condensation reactions in aqueous-organic solvent. Biocat. Biotrans. 14:205–217

Kuwajima K (1996) The molten globule state of α-lactalbumin. FASEB J. 10:102–109

Kuwajima K, Yamaya H, Miwa S, Sugai S, Nagamura T (1987) Rapid formation of secondary structure framework in protein folding studied by stopped-flow circular dichroism. FEBS Lett. 221:115–118

Kuwajima K, Semisotnov GV, Finkelstein AV, Sugai S, Ptitsyn OB (1993) Secondary structure of globular proteins at the early and the final stages in protein folding. FEBS Lett. 334:265–268

Kuwajima K, Yamaya H, Sugai S (1996) The burst-phase intermediate in the refolding of β-lactoglobulin studied by stopped-flow circular dichroism and absorption spectroscopy. J. Mol. Biol. 264:806–822

Ladurner AG, Itzhaki LS, dePratGay G, Fersht AR (1997) Complementation of peptide fragments of the single domain protein chymotrypsin inhibitor 2. J. Mol. Biol. 273:317–329

Leeson DT, Gai F, Rodriguez HM, Gregoret LM, Dyer RB (2000) Protein folding and unfolding on a complex energy landscape. Proc. Natl. Acad. Sci. USA 97:2527–2532

Levinthal C (1968) Are there pathways for protein folding? J. Chim. Phys. 85:44–45

Lewis JW, Tilton RF, Einterz CM, Milder SJ, Kuntz ID, Kliger DS (1985) New Technique for measuring circular dichroism changes on a nanosecond time scale. Application to carbonmonoxy myoglobin and carbonmonoxy hemoglobin. J. Phys. Chem. 89:289–294

Lewis JW, Yee GG, Kliger DS (1987) Implementation of an optical multichannel analyzer for nanosecond flash photolysis measurements. Rev. Sci. Instrum. 58:939–944

Lewis JW, Goldbeck RA, Kliger DS, Xie XL, Dunn RC, Simon JD (1992) Time-resolved circular dichroism spectroscopy: Experiment, theory, and applications to biological systems. J. Phys. Chem. 96:5243–5254

Lide DR (ed) (1993) CRC handbook of chemistry and physics. CRC Press, Boca Raton Ann Arbor London Tokyo, 74th Ed.

Lin SH, Cheung HC (1992) The kinetics of a two-state transition of myosin subfragment 1: A temperature-jump relaxation study. FEBS Lett. 304:184–186

Lin Y, Gerfen GJ, Rousseau DL, Yeh SR (2003) Ultrafast microfluidic mixer and freeze-quenching device. Anal Chem. 75:5381–5386

López-Hernández E, Serrano L (1996) Structure of the transition state for folding of the 129-aa protein CheY resembles that of a smaller protein, CI2. Folding & Design 1:43–55

Lubienski MJ, Bycroft M, Freund SMV, Fersht AR (1994) Three-dimensional solution structure and ^{13}C-assignments of barstar using nuclear magnetic resonance spectroscopy. Biochemistry 33:8866–8877

Luchins J, Beychok S (1978) Far-ultraviolet stopped-flow circular dichroism. Science 199:425–426

Luo YZ, Baldwin RL (1998) Trifluoroethanol stabilizes the pH 4 folding intermediate of sperm whale apomyoglobin. J. Mol. Biol. 279:49–57

Luthardt G, Frömmel C (1994) Local polarity analysis: A sensitive method that discriminates between native proteins and incorrectly folded models. Protein Engineering 7:627–631

Makhatadze GI, Privalov PL (1993) Contribution of hydration to protein folding thermodynamics. 1. The enthalpy of hydration. J. Mol. Biol. 232:639–659

Maness SJ, Franzen S, Gibbs AC, Causgrove TP, Dyer RB (2003) Nanosecond temperature jump relaxation dynamics of cyclic β-hairpin peptides. Biophys. J. 84:3874–3882

Martin J, Mayhew M, Langer T, Hartl FU (1993) The reaction cycle of GroEL and GroES in chaperonin-assisted protein folding. Nature 366:228–233

Martin J, Goldie KN, Engel A, Hartl FU (1994) Topology of the morphological domains of the chaperonin GroEL visualized by immunoelectron microscopy. Biol. Chem. Hoppe–Seyler 375:635–639

Masters CL, Beyreuther K (1997) Tracking turncoat prion proteins. Nature 388:228–229

Mateu MG, DelPino MMS, Fersht AR (1999) Mechanism of folding and assembly of a small tetrameric protein domain from tumor suppressor p53. Nature Struct. Biol. 6:191–198

Matouschek A, Fersht AR (1991) Protein engineering in analysis of protein folding pathways and stability. Methods Enzymol. 202:82–112

Matouschek A, Fersht AR (1993) Application of physical organic chemistry to engineered mutants of proteins: Hammond postulate behavior in the transition state of protein folding. Proc. Natl. Acad. Sci. USA 90:7814–7818

Matouschek A, Kellis JT, Serrano L, Fersht AR (1989) Mapping the transition state and pathway of protein folding by protein engineering. Nature 340:122–126

Matouschek A, Kellis JT, Serrano L, Bycroft M, Fersht AR (1990) Transient folding intermediates characterized by protein engineering. Nature 346:440–445

Matouschek A, Serrano L, Fersht AR (1992) The folding of an enzyme. IV. Structure of an intermediate in the refolding of barnase analysed by a protein engineering procedure. J. Mol. Biol. 224:819–835

Matthew JB (1985) Electrostatic effects in proteins. Annu. Rev. Biophys. Biophys. Chem. 14:387–417

Mayor U, Johnson CM, Daggett V, Fersht AR (2000) Protein folding and unfolding in microseconds to nanoseconds by experiment and simulation. Proc. Natl. Acad. Sci. USA 97:13518–13522

Mayor U, Guydosh NR, Johnson CM, Grossmann JG, Sato S, Jas GS, Freund SM, Alonso DO, Daggett V, Fersht AR (2003) The complete folding pathway of a protein from nanoseconds to microseconds. Nature 421:863–867

McCaldon P, Argos P (1988) Oligopeptide biases in protein sequences and their use in predicting protein coding regions in nucleotide sequences. Proteins: Struct., Funct., Genetics 4:99–122

McMaster TJ, Miles MJ, Walsby AE (1996) Direct observation of protein secondary structure in gas vesicles by atomic force microscopy. Biophys. J. 70:2432–2436

Merritt EA, Murphy MEP (1994) Raster3D version 2.0: A program for photorealistic molecular graphics. Acta Cryst. D50:869–873

Michels PC, Hei D, Clark DS (1996) Pressure effects on enzymatic activity and stability at high temperatures. Adv. Prot. Chem. 48:341–376

Michnick SW, Shakhnovich E (1998) A strategy for detecting the conservation of folding-nucleus residues in protein superfamilies. Folding & Design 3:239–251

Milder SJ, Björling SC, Kuntz ID, Kliger DS (1988) Time-resolved circular dichroism and absorption studies on the photolysis reaction of carbonmonoxy myoglobin. Biophys. J. 53:659–664

Milla ME, Brown BM, Waldburger CD, Sauer RT (1995) P22 Arc repressor: transition state properties inferred from mutational effects on the rates of protein unfolding and refolding. Biochemistry 34:13914–13919

Mines GA, Pascher T, Lee SC, Winkler JR, Gray HB (1996) Cytochrome c folding triggered by electron transfer. Chemistry & Biology 3:491–497

Miranker A, Kruppa GH, Robinson CV, Aplin RT, Dobson CM (1996a) Isotope-labeling strategy for the assignment of protein fragments generated for mass spectrometry. J. Am. Chem. Soc. 118:7402–7403

Miranker A, Robinson CV, Radford SE, Dobson CM (1996b) Investigation of protein folding by mass spectrometry. FASEB J. 10:93–101

Mirny LA, Abkevich VI, Shakhnovich EI (1998) How evolution makes proteins fold quickly. Proc. Natl. Acad. Sci. USA 95:4976–4981

Mozhaev VV, Heremans K, Frank J, Masson P, Balny C (1996) High-pressure effects on protein structure and function. Proteins: Struct., Funct., Genetics 24:81–91

Muñoz V, Thompson PA, Hofrichter J, Eaton WA (1997) Folding dynamics and mechanism of β-hairpin formation. Nature 390:196–199

Muñoz V, Henry ER, Hofrichter J, Eaton WA (1998) A statistical-mechanical model for β-hairpin kinetics. Proc. Natl. Acad. Sci. USA 95:5872–5879

Murzin AG, Brenner SE, Hubbard T, Chothia C (1995) SCOP: A structural classification of proteins database for the investigation of sequences and structures. J. Mol. Biol. 247:536–540

Myers JK, Oas TG (2002) Mechanism of fast protein folding. Annu. Rev. Biochem. 71:783–815

Nakanishi K, Berova N, Woody RW (eds) (1994) Circular Dichroism. VCH, New York Weinheim Cambridge

Narasimhulu S (1993) Substrate induced spin-state transition in cytochrome P-450LM2: A temperature-jump relaxation study. Biochemistry 32:10344–10350

Nath U, Udgaonkar JB (1997a) How do proteins fold? Current Science 72:180–191

Nath U, Udgaonkar JB (1997b) Folding of tryptophan mutants of barstar: Evidence for an initial hydrophobic collapse on the folding pathway. Biochemistry 36:8602–8610

Neira JL, Fersht AR (1999) Exploring the folding funnel of a polypeptide chain by biophysical studies on protein fragments. J. Mol. Biol. 285:1309–1333

Nishi I, Kataoka N, Tokunaga F, Goto Y (1994) Cold denaturation of the molten globule states of apomyoglobin and a profile for protein folding. Biochemistry 33:4903–4909

Nölting B (1991) A new spectrometer for the simultaneous measurement of absorption and circular dichroism (in German). Ph.D. thesis, University of Bochum.

Nölting B (1995) Relation between adiabatic and pseudoadiabatic compressibility in ultrasonic velocimetry. J. theor. Biol. 175:191–196

Nölting B (1996) Temperature-jump induced fast refolding of cold-unfolded protein. Biochem. Biophys. Res. Comm. 227:903–908

Nölting B (1998a) Structural resolution of the folding pathway of a protein by correlation of Φ-values with inter-residue contacts. J. theor. Biol. 194:419–428

Nölting B (1998b) The distribution of temperature in globular molecules, cells, or droplets in temperature-jump-, sound velocity-, and pulsed LASER-experiments. J. Phys. Chem. B. 102:7506–7509

Nölting B (1999) Analysis of the folding pathway of chymotrypsin inhibitor by correlation of Φ-values with inter-residue contacts. J. theor. Biol. 197:113–121

Nölting B (2003) Methods in Modern Biophysics. Springer-Verlag, Berlin Heidelberg New York

Nölting B, Sligar SG (1993) Adiabatic compressibility of molten globules. Biochemistry 32:12319–12323

Nölting B, Andert K (2000) Mechanism of protein folding. Proteins: Struct. Funct. Genetics 41:288–298

Nölting B, Jung C, Snatzke G (1992) Multichannel circular dichroism investigations of the structural stability of bacterial cytochrome P-450. Biochim. Biophys. Acta 1100:171–176

Nölting B, Jiang M, Sligar SG (1993) The acidic molten globule state of α-lactalbumin probed by sound velocity. J. Am. Chem. Soc. 115:9879–9882

Nölting B, Golbik R, Fersht AR (1995) Submillisecond events in protein folding. Proc. Natl. Acad. Sci. USA 92:10668–10672

Nölting B, Golbik R, Neira JL, Soler-González AS, Schreiber G, Fersht AR (1997a) The folding pathway of a protein at high resolution from microseconds to seconds. Proc. Natl. Acad. Sci. USA 94:826–830

Nölting B, Golbik R, Soler-González AS, Fersht AR (1997b) Circular dichroism of denatured barstar suggests residual structure. Biochemistry 36:9899–9905

Nölting B, Schälike W, Hampel P, Grundig F, Gantert S, Sips N, Bandlow W, Qi PX (2003) Structural determinants of the rate of protein folding. J. theor. Biol. 223:299–307

Nölting B, Jülich D, Vonau W, Andert K (2004) Evolutionary computer programming of protein folding and structure predictions. J. theor. Biol. 229:13–18

Nymeyer H, Garcia AE, Onuchic JN (1998) Folding funnels and frustration in off-lattice minimalist protein landscapes. Proc. Natl. Acad. Sci. USA 95:5921–5928

O'Connor DB, Goldbeck RA, Hazzard JH, Kliger DS, Cusanovich MA (1993) Time-resolved absorption and magnetic circular dichroism spectroscopy of cytochrome c3 from *Desulfovibrio*. Biophys. J. 65:1718–1726

O'Neil KT, DeGrado WF (1990) A thermodynamic scale for the helix-forming tendencies of the commonly occurring amino acids. Science 250:646–651

Oliveberg M, Fersht AR (1996a) Formation of electrostatic interactions on the protein folding pathway. Biochemistry 35:2726–2737

Oliveberg M, Fersht AR (1996b) Thermodynamics of transient conformations in the folding pathway of barnase: Reorganization of the folding intermediate at low pH. Biochemistry 35:2738–2749

Onuchic JN, Wolynes PG, Luthey-Schulten Z, Socci ND (1995) Toward an outline of the topography of a realistic protein folding funnel. Proc. Natl. Acad. Sci. USA 92:3626–3630

Onuchic JN, Socci ND, Luthey-Schulten Z, Wolynes PG (1996) Protein folding funnels: The nature of the transition state ensemble. Folding & Design 1:441–450

Otzen DE, Fersht AR (1998) Folding of circular and permuted chymotrypsin inhibitor 2: retention of the folding nucleus. Biochemistry 37:8139–8146

Otzen DE, Itzhaki LS, ElMasry NF, Jackson SE, Fersht AR (1994) Structure of the transition state of the folding/unfolding of the barley chymotrypsin inhibitor 2 and its implications for mechanisms of protein folding. Proc. Natl. Acad. Sci. USA 91:10422–10425

Pabit SA, Roder H, Hagen SJ (2004) Internal friction controls the speed of protein folding from a compact configuration. Biochemistry 43:12532–12538

Pappu RV, Weaver DL (1998) The early folding kinetics of apomyoglobin. Protein Sci. 7:480–490

Park SH, O'Neil KT, Roder H (1997) An early intermediate in the folding reaction of the B1 domain of protein G contains a native-like core. Biochemistry 36:14277–14283

Pascher T, Chesick JP, Winkler JR, Gray HB (1996) Protein folding triggered by electron transfer. Science 271:1558–1560

Petrich JW, Martin JL, Houde D, Poyart C, Orszag A (1987) Time-resolved Raman spectroscopy with subpicosecond resolution: Vibrational cooling and delocalization of strain energy in photodissociated carbonmonoxy hemoglobin. Biochemistry 26:7914–7923

Pfeil W (1981) Thermodynamics of α-lactalbumin unfolding. Biophys. Chem. 13:181–186

Pfeil W (1988) Protein unfolding. In: Jones MN (ed) Biochemical thermodynamics. Elsevier, Amsterdam, 53–99

Pfeil W (1993) Thermodynamics of apocytochrome b5 unfolding. Protein Science 2:1497–1501

Pfeil W, Welfle K, Bychkova VE (1991) Guanidine-hydrochloride titration of the unfolded apocytochrome c studied by calorimetry. Studia Biophysica 140:5–12

Pfeil W, Nölting B, Jung C (1993a) Apocytochrome P-450cam is a native protein with some intermediate-like properties. Biochemistry 32:8856–8862

Pfeil W, Nölting B, Jung C (1993b) Thermodynamic properties of apocytochrome P-450cam. In: Tweel WJJvd, Harder A, Buitelaar RM (eds) Stability and Stabilization of Enzymes. Elsevier, Amsterdam, 407–414

Pflumm M, Luchins J, Beychok S (1986) Stopped-flow circular dichroism. Methods Enzymol. 130:519–534

Phillips CM, Mizutani Y, Hochstrasser RM (1995) Ultrafast thermally-induced unfolding of ribonuclease A. Proc. Natl. Acad. Sci. USA 92:7292–7296

Plaxco KW, Dobson CM (1996) Time-resolved biophysical methods in the study of protein folding. Curr. Opin. Struct. Biol. 6:630–636

Plaxco KW, Simons KT, Baker DJ (1998) Contact order, transition state placement and the refolding rates of single-domain proteins. J. Mol. Biol. 277:985–994

Porschke D (1996) Analysis of chemical and physical relaxation processes of polyelectrolytes by electric field pulse methods. A comparison of critical comments with facts. Ber. Bunsen Gesellschaft - Phys. Chem. Chem. Phys. 100:715–720

Porschke D, Obst A (1991) An electric-field-jump apparatus with ns time resolution for electro-optical measurements at physiological salt concentrations. Rev. Sci. Instrum. 62:818–820

Poulos TL, Finzel BC, Howard, AJ (1986) Crystal structure of substrate-free *Pseudomonas putida* cytochrome P-450. Biochemistry 25:5314–5322

Presta LG, Rose GD (1988) Helix signal in proteins. Science 240:1632–1641

Prince SM, Papiz MZ, Freer AA, McDermott G, Hawthornthwaite-Lawless AM, Cogdell RJ, Isaacs NW (1997) Apoprotein structure in the LH2 complex from *Rhodopseudomonas acidophila* strain 10050: Modular assembly and protein–pigment interactions. J. Mol. Biol. 268:412–423

Privalov PL (1979) Stability of protein: Small globular proteins. Adv. Prot. Chem. 33:167–241

Privalov PL (1990) Cold denaturation of proteins. Crit. Rev. Biochem. Mol. Biol. 25:281–305

Privalov PL (1996) Intermediate states in protein folding. J. Mol. Biol. 258:707–725

Privalov PL, Makhatadze GI (1993) Contribution of hydration to protein folding thermodynamics. 2. The entropy and Gibbs energy of hydration. J. Mol. Biol. 232:660–679

Pryse KM, Bruckman TG, Maxfield BW, Elson EL (1992) Kinetics and mechanism of the folding of cytochrome c. Biochemistry 31:5127–5136

Ptitsyn OB (1981) Protein folding: General physical model. FEBS Lett. 131:197–202

Ptitsyn OB (1994) Kinetic and equilibrium intermediates in protein folding. Protein Eng. 7:593–596

Ptitsyn OB (1995) Molten globule and protein folding. Adv. Prot. Chem. 47:83–229

Ptitsyn OB (1998) Protein folding: Nucleation and compact intermediates. Biochemistry (Moscow) 63:367–373

Ptitsyn OB, Rashin AA (1975) A model of myoglobin self-organization. Biophys. Chem. 3:1–20

Qi PX, Beckman RA, Wand AJ (1996) Solution structure of horse heart ferricytochrome c and detection of redox-related structural changes by high-resolution ^{1}H NMR. Biochemistry 35:12275–12287

Qiu L, Pabit SA, Roitberg AE, Hagen SJ (2002) Smaller and faster: the 20-residue Trp-cage protein folds in 4 µs. J. Am. Chem. Soc. 124:12952–12953

Rackovsky S, Scheraga HA (1977) Hydrophobicity, hydrophilicity, and the radial and orientational distributions of residues in native proteins. Proc. Natl. Acad. Sci. USA 74:5248–5251

Radzicka A, Wolfenden R (1988) Comparing the polarities of the amino acids: Side chain distribution coefficients between the vapor-phase, cyclohexane, 1-octanol, and neutral aqueous solution. Biochemistry 27:1664–1670

Regenfuss P, Clegg RM (1987) Diffusion-controlled association of a dye, 1-anilinonaphthalene-8-sulfonic acid, to a protein, bovine serum albumin, using a fast-flow microsecond mixer and stopped-flow. Biophys. Chem. 26:83–89

Regenfuss P, Clegg RM, Fulwyler MJ, Barrantes FJ, Jovin TM (1985) Mixing liquids in microseconds. Rev. Sci. Instrum. 56:283–290

Richards FM (1974) The interpretation of protein structures: Total volume, group volume distributions and packing density. J. Mol. Biol. 82:1–14

Richardson JS, Richardson DC (1988) Amino acid preferences for specific locations at the ends of α-helices. Science 240:1648–1652

Riddle DS, Grantcharova VP, Santiago JV, Alm E, Ruczinski I, Baker D (1999) Experiment and theory highlight role of native state topology in SH3 folding. Nature Struct. Biol. 6:1016–1024

Riek R, Hornemann S, Wider G, Billeter M, Glockshuber R, Wüthrich K (1996) NMR structure of the mouse prion protein domain PrP(121–231) and inherited human prion diseases. Nature 382:180–182

Robinson CR, Sauer RT (1996) Equilibrium stability and submillisecond refolding of a designed single-chain Arc repressor. Biochemistry 35:13878–13884

Roder H, Colon W (1997) Kinetic role of early intermediates in protein folding. Curr. Opin. Struct. Biol. 7:15–28

Roder H, Elöve GA, Englander SW (1988) Structural characterization of folding intermediates in cytochrome c by H-exchange labeling and proton NMR. Nature 335:700–704

Rose GD (1987) Protein hydrophobicity: Is it the sum of its parts? Proteins: Struct., Funct., Genetics 2:79–80

Sabelko J, Erwin J, Gruebele M (1998) Cold-denatured ensemble of apo-myoglobin: Implications for the early steps of folding. J. Phys. Chem. B 102:1806–1819

Sarvazyan AP (1991) Ultrasonic velocimetry of biological compounds. Annu. Rev. Biophys. Biophys. Chem. 20:321–342

Sauder JM, Mackenzie NE, Roder H (1996) Kinetic mechanism of folding and unfolding of *Rhodobacter capsulatus* cytochrome c2. Biochemistry 35:16852–16862

Scheraga HA (1996) Recent developments in the theory of protein folding: Searching for the global energy minimum. Biophys. Chem. 59:329–339

Schindler T, Herrler M, Marahiel MA, Schmid F-X (1995) Extremely rapid protein folding in the absence of intermediates. Nat. Struct. Biol. 2:663–673

Schreiber G, Fersht AR (1993a) Interaction of barnase with its polypeptide inhibitor barstar studied by protein engineering. Biochemistry 32:5145–5150

Schreiber G, Fersht AR (1993b) The refolding of *cis*-peptidylprolyl and *trans*-peptidylprolyl isomers of barstar. Biochemistry 32:11195–11203

Schreiber G, Fersht AR (1995) Energetics of the protein–protein interactions: Analysis of the barnase–barstar interface by single mutations and double mutant cycles. J. Mol. Biol. 248:478–486

Schreiber G, Buckle AM, Fersht AR (1994) Stability and function: Two constraints in the evolution of barstar and other proteins. Structure 2:945–951

Schwarz G, Seelig J (1968) Kinetic properties and the electric field effect of the helix–coil transition of poly(benzyl-L-glutamate) determined from dielectric relaxation measurements. Biopolymers 6:1263–1277

Semisotnov GV, Kihara H, Kotova NV, Kimura K, Amemiya Y, Wakabayashi K, Serdyuk IN, Timchenko AA, Chiba K, Nikaido K, Ikura T, Kuwajima K (1996) Protein globularization during folding: A study by synchrotron small-angle X-ray scattering. J. Mol. Biol. 262:559–574

Serrano L, Kellis JT Jr, Cann P, Matouschek A, Fersht AR (1992a) The folding of an enzyme. II. Substructure of barnase and the contribution of different interactions to protein stability. J. Mol. Biol. 224:783–804

Serrano L, Matouschek A, Fersht AR (1992b) The folding of an enzyme. III. Structure of the transition state for unfolding of barnase analyzed by a protein engineering procedure. J. Mol. Biol. 224:805–818

Serrano L, Matouschek A, Fersht AR (1992c) The folding of an enzyme. VI. The folding pathway of barnase: Comparison with theoretical models. J. Mol. Biol. 224:847–859

Service RF (1996) Folding proteins caught in the act. Science 273:29–30

Shakhnovich EI (1997) Theoretical studies of protein folding thermodynamics and kinetics. Cur. Opin. Struct. Biol. 7:29–40

Shakhnovich EI, Abkevich V, Ptitsyn O (1996) Conserved residues and the mechanism of protein folding. Nature 379:96–98

Shapiro DB, Goldbeck RA, Che DP, Esquerra RM, Paquette SJ, Kliger DS (1995) Nanosecond optical rotatory dispersion spectroscopy: Application to photolyzed hemoglobin–CO kinetics. Biophys. J. 68:326–334

Shastry MCR, Udgaonkar JB (1995) The folding mechanism of barstar: Evidence for multiple pathways and multiple intermediates. J. Mol. Biol. 247:1013–1027

Shastry MCR, Roder H (1998) Evidence for barrier-limited protein folding kinetics on the microsecond time scale. Nat. Struct. Biol. 5:385–392

Shastry MCR, Luck SD, Roder H (1998) A continuous-flow capillary mixing method to monitor reactions on the microsecond time scale. Biophys. J. 74:2714–2721

Shoemaker BA, Wang J, Wolynes PG (1997) Structural correlations in protein folding funnels. Proc. Natl. Acad. Sci. USA 94:777–782

Shoemaker BA, Wang J, Wolynes PG (1999) Exploring structures in protein folding funnels with free energy functionals: the transition state ensemble. J. Mol. Biol. 287:675–694

Shortle D, Simons KT, Baker D (1998) Clustering of low-energy conformations near the native structures of small proteins. Proc. Natl. Acad. Sci. USA 95:11158–11162

Sligar SG (1976) Coupling of spin, substrate, and redox equilibria in cytochrome P-450. Biochemistry 15:5399–5406

Sohl JL, Jaswal SS, Agard DA (1998) Unfolded conformations of α-lytic protease are more stable than its native state. Nature 395:817–819

Sosnick TR, Mayne L, Hiller R, Englander SW (1994) The barriers in protein folding. Nat. Struct. Biol. 1:149–156

Sosnick TR, Mayne L, Englander SW (1996) Molecular collapse: The rate-limiting step in two-state cytochrome c folding. Proteins: Struct. Funct. Genetics 24:413–426

Stark B, Nölting B, Jahn H, Andert K (1992) Method for determining the electron number in charge-coupled measurement devices. Optical Engineering 31:852–856

Steinhoff HJ, Lieutenant K, Redhardt A (1989) Conformational transition of aquomethemoglobin: Intramolecular histidine E7 binding reaction to the heme iron in the temperature range between 220 K and 295 K as seen by EPR and temperature-jump measurements. Biochim. Biophys. Acta 996:49–56

Stouten PFW, Frömmel C, Nakamura H, Sander C (1993) An effective solvation term based on atomic occupancies for use in protein simulations. Molecular Simulation 10:2–6

Takahashi MT, Alberty RA (1969) The pressure-jump method. Methods Enzymol. 16:31–55

Takahashi S, Yeh SR, Das TK, Chan CK, Gottfried DS, Rousseau DL (1997) Folding of cytochrome c initiated by submillisecond mixing. Nat. Struct. Biol. 4:44–50

Tamura Y, Gekko K (1995) Compactness of thermally and chemically denatured ribonuclease A as revealed by volume and compressibility. Biochemistry 34:1878–1884

Tanaka N, Kunugi S (1996) Effect of pressure on the deuterium-exchange of α-lactalbumin and β-lactoglobulin. Int. J. Biol. Macromol. 18:33–39

Teilum K, Maki K, Kragelund BB, Poulsen FM, Roder H (2002) Early kinetic intermediate in the folding of acyl-CoA binding protein detected by fluorescence labeling and ultrarapid mixing. Proc. Natl. Acad. Sci. USA 99:9807–9812

Thompson PA (1997) Laser temperature-jump for the study of early events in protein folding. In: Marshak DR (ed) Techniques in Protein Chemistry VIII. San Diego: Academic Press.

Thompson PA, Eaton WA, Hofrichter J (1997) Laser temperature-jump study of the helix–coil kinetics of an alanine peptide interpreted with a 'kinetic zipper' model. Biochemistry 36:9200–9210

Thornton JM, Jones DT, MacArthur MW, Orengo CM, Swindells MB (1995) Protein folds: Towards understanding folding from inspection of native structures. Phil. Transactions Royal Soc. London B 348:71–79

Tilton RF Jr, Dewan JC, Petsko GA (1992) Effects of temperature on protein structure and dynamics: X-ray crystallographic studies of the protein ribonuclease A at nine different temperatures from 98 to 320 K. Biochemistry 31:2469–2481

Topchieva IN, Sorokina EM, Kurganov BI, Zhulin VM, Makarova ZG (1996) High pressure induced complexes of α-chymotrypsin with block-copolymers based on ethylene and propylene oxides. Biochemistry (Moscow) 61:746–749

Tsong TY (1982) Viscosity-dependent conformational relaxation of ribonuclease A in the thermal unfolding zone. Biochemistry 21:1493–1498

Tsong TY, Baldwin RL, Elson EL (1971) The sequential unfolding of ribonuclease A: Detection of a fast initial phase in the kinetics of unfolding. Proc. Natl. Acad. Sci. USA 68:2712–2715

Udgaonkar JB, Baldwin RL (1988) NMR evidence for an early framework intermediate on the folding pathway of ribonuclease A. Nature 335:694–699

Urbanke C, Wray J (2001) A fluorescence temperature-jump study of conformational transitions in myosin subfragment 1. Biochem. J. 358:165–173

Varotsis C, Babcock GT (1993) Nanosecond time-resolved resonance Raman spectroscopy. Methods Enzymol. 226:409–431

Velluz L, Legrand M, Grosjean M (1965) Optical circular dichroism. Verlag Chemie, Weinheim

Viguera AR, Villegas V, Aviles FX, Serrano L (1997) Favorable native-like helical local interactions can accelerate protein folding. Folding & Design 2:23–33

Vu DM, Myers JK, Oas TG, Dyer RB (2004) Probing the folding and unfolding dynamics of secondary and tertiary structures in a three-helix bundle protein. Biochemistry 43: 3582–3589

Wallenhorst WF, Green SM, Roder H (1997) Kinetic evidence for folding and unfolding intermediates in staphylococcal nuclease. Biochemistry 36:5795–5805

Walz FG (1992) Relaxation kinetics of ribonuclease T1 binding with guanosine and 3'-GMP. Biochim. Biophys. Acta 1159:327–334

Wang MS, Gandour RD, Rodgers J, Haslam JL, Schowen RL (1975) Transition-state structure for a conformation change of ribonuclease. Bioorg. Chem. 4:392–406

Wangikar PP, Michels PC, Clark DS, Dordick JS (1997) Structure and function of subtilisin BPN' solubilized in organic solvents. J. Am. Chem. Soc. 119:70–76

Warshel A, Levitt M (1976) Theoretical studies of enzymatic reactions: Dielectric, electrostatic and steric stabilization of the carbonium ion in the reaction of lysozyme. J. Mol. Biol. 103:227–249

Weber G (1993) Thermodynamics of the association and the pressure dissociation of oligomeric proteins. J. Phys. Chem. 97:7108–7115

Weber G (1996) Persistent confusion of total entropy and chemical-system entropy in chemical thermodynamics. Proc. Natl. Acad. Sci. USA 93:7452–7453

Wen YX, Chen EF, Lewis JW, Kliger DS (1996) Nanosecond time-resolved circular dichroism measurements using an up-converted Ti-sapphire LASER. Rev. Sci. Instrum. 67:3010–3016

Wetlaufer DB (1962) Ultraviolet spectra of proteins and amino acids. Adv. Protein Chem. 17:303–390

Wetlaufer DB (1973) Nucleation, rapid folding, and globular intrachain regions in proteins. Proc. Natl. Acad. Sci. USA 70:697–701

Wetlaufer DB (1990) Nucleation in protein folding: Confusion of structure and process. Trends in Biochemical Sciences 15:414–415

Wetlaufer DB, Xie Y (1995) Control of aggregation in protein refolding: A variety of surfactants promote renaturation of carbonic anhydrase II. Protein Sci. 4:1535–1543

Williams DH, Fleming I (1995) Spectroscopic methods in organic chemistry. McGraw-Hill, London

Williams S, Causgrove TP, Gilmanshin R, Fang KS, Callender RH, Woodruff WH, Dyer RB (1996) Fast events in protein folding: Helix melting and formation in a small peptide. Biochemistry 35:691–697

Wilson CJ, Wittung-Stafshede P (2005) Role of structural determinants in folding of the sandwich-like protein *Pseudomonas aeruginosa* azurin. Proc. Natl. Acad. Sci. USA 102:3984–3987

Wittung-Stafshede P, Lee JC, Winkler JR, Gray HB (1999) Cytochrome b_{562} folding triggered by electron transfer: approaching the speed limit for formation of a four-helix-bundle protein. Proc. Natl. Acad. Sci. USA 96:6587–6590

Wolynes PG (1996) Symmetry and the energy landscapes of biomolecules. Proc. Natl. Acad. Sci. USA 93:14249–14255

Wolynes PG, Onuchic JN, Thirumalai D (1995) Navigating the folding routes. Science 267:1619–1620

Wolynes PG, Luthey-Schulten Z, Onuchic JN (1996) Fast folding experiments and the topography of protein folding energy landscapes. Chem. & Biol. 3:425–432

Wong KB, Freund SMV, Fersht AR (1996) Cold denaturation of barstar: ^{1}H, ^{15}N and ^{13}C NMR assignment and characterization of residual structure. J. Mol. Biol. 259:805–818

Woodruff WH, Einarsdóttir O, Dyer RB, Bagley KA, Palmer G, Atherton SJ, Goldbeck RA, Dawes TD, Kliger DS (1991) Nature and functional implications of the cytochrome a3 transients after photodissociation of CO-cytochrome oxydase. Proc. Natl. Acad. Sci. USA 88:2588–2592

Wüthrich K (1986) NMR of proteins and nucleic acids. Wiley, New York

Xie X, Simon JD (1989) Picosecond time-resolved circular dichroism spectroscopy: Experimental details and applications. Rev. Sci. Instrum. 60:2614–2627

Xie X, Simon JD (1991) Protein conformational relaxation following photodissociation of CO from carbonmonoxy myoglobin: Picosecond circular dichroism and absorption studies. Biochemistry 30:3682–3692

Xu Y, Oyola R, Gai F (2003) Infrared study of the stability and folding kinetics of a 15-residue β-hairpin. J. Am. Chem. Soc. 125:15388–15394

Yamamoto K, Mizutani Y, Kitagawa T (2000) Nanosecond temperature jump and time-resolved Raman study of thermal unfolding of ribonuclease A. Biophys. J. 79:485–495

Yanagawa H, Yoshida K, Torigoe C, Park JS, Sato K, Shirai T, Go M (1993) Protein anatomy: Functional roles of barnase module. J. Biol. Chem. 268:5861–5865

Yang JT, Wu CSC, Martinez HM (1986) Calculation of protein conformation from circular dichroism. Methods Enzymol. 130:208–269

Yeh SR, Rousseau DL (1998) Folding intermediates in cytochrome c. Nat. Struct. Biol. 5:222–228

Yeh SR, Takahashi S, Fan BC, Rousseau DL (1997) Ligand exchange during cytochrome c folding. Nat. Struct. Biol. 4:51–56

Yuzawa T, Kato C, George MW, Hamaguchi HO (1994) Nanosecond time-resolved infrared spectroscopy with a dispersive scanning spectrometer. Appl. Spectroscopy 48:684–690

Zhang CF, Lewis JW, Cerpa R, Kuntz ID, Kliger DS (1993) Nanosecond circular dichroism spectral measurements: Extension to the far-ultraviolet region. J. Phys. Chem. 97:5499–5505

Zhu X, Zhao X, Burkholder WF, Gragerov A, Ogata CM, Gottesman ME, Hendrickson, WA (1996) Structural analysis of substrate binding by the molecular chaperone DnaK. Science 272:1606–1614

Zhu Y, Alonso DO, Maki K, Huang CY, Lahr SJ, Daggett V, Roder H, DeGrado WF, Gai F
(2003) Ultrafast folding of α_3D: a de novo designed three-helix bundle protein.
Proc. Natl. Acad. Sci. USA 100:15486–15491

Zhu YJ, Fu XR, Wang T, Tamura A, Takada S, Savan JG, Gai F (2004) Guiding the search for a
protein's maximum rate of folding. Chem. Phys. 307:99–109

Zubay G (1993) Biochemistry. Wm. C. Brown Publishers, Dubuque Melbourne Oxford, 3rd Ed.

Index